An Engineering Data Book

Edited by

A. J. Munday
and
R. A. Farra

Published by
THE MACMILLAN PRESS LTD
London and Basingstoke

ISBN 0 333 25829 0

Printed in Hong Kong

M

First edition 1979
Reprinted 1979, 1980, 1982, 1983

1. UNITS AND ABBREVIATIONS

1.1 Decimal prefixes

symbol	prefix	factor by which unit is multiplied
T	tera	10^{12}
G	giga	10^{9}
M	mega	10^{6}
k	kilo	10^{3}
h	hecto	10^{2}
da	deca	10
d	deci	10^{-1}
c	centi	10^{-2}
m	milli	10^{-3}
μ	micro	10^{-6}
n	nano	10^{-9}
p	pico	10^{-12}

1.2 SI units

(i) Basic units

unit symbol	unit	quantity
m	metre	length
kg	kilogramme	mass
s	second	time
A	ampere	electric current
K	kelvin	thermodynamic temperature
cd	candela	luminous intensity

(ii) Supplementary and derived units

quantity	unit	symbol	equivalent
plane angle	radian	rad	-
force	newton	N	kg m/s^2
work, energy heat	joule	J	N m
power	watt	W	J/s
frequency	hertz	Hz	s^{-1}
viscosity:			
kinematic		m^2/s	10^6 cSt (centi-stoke)
dynamic		Ns/m^2=Pa s	10^3 cP (centi-poise)
pressure		Pa=N/m^2	Called pascal, Pa
stress		Pa or N/m^2	
	electrical units		
potential	volt	V	W/A
resistance	ohm	Ω	V/A
charge	coulomb	C	A s
capacitance	farad	F	A s/V
electric field strength	-	V/m	
electric flux density	-	C/m^2	
	magnetic units		
magnetic flux	weber	Wb	V s = Nm/A
inductance	henry	H	V s/A = Nm/A^2
magnetic field strength	-	A/m	
magnetic flux density	tesla	T	Wb/m^2 = N/(Am)

1.3 Conversion factors for other units into SI units

Length, area, volume

1 in = 25.4 mm exactly   1 Å = 10^{-10} m

1 ft = 0.3048 m   1 thou=1 mil = 0.001 in = 25.4 μm

1 yd = 0.914 m   1 micron = 1μm

1 mile = 5280 ft = 1.609 km

1 acre = 0.4047 ha (Hectare) = 4047 m^2

1 in^3 = 16.39 cm^3

1 ft^3 = 0.02832 m^3 1 nm = 1852 m

1 gal = 0.1605 ft^3 = 4546 cm^3 = 4.546 ℓ (Litre)

1 USgal = 0.1337 ft^3 = 3785 cm^3

Velocity

1 mile/h = 1.467 ft/s = 1.609 km/h = 0.447 m/s

1 knot = 1.689 ft/s = 1.853 km/h = ~~0.514~~ m/s 0.5148 ~~0.3575~~ m/s

Mass

1 lb = 0.4536 kg

1 slug = 32.17 lb = 14.59 kg

1 ton = 2240 lb = 1016 kg

1 tonne = 1 Mg = 1 metric ton

Flowrate

1 ft^3/s (1 cusec) = 0.02832 m^3/s

1 gal/min = 7.577 10^{-5} m^3/s = 0.07577 dm^3/s

Density

1 lb/in^3 = 27.68 g/cm^3

1 lb/ft^3 = 16.02 kg/m^3

1 slug/ft^3 = 515.4 kg/m^3

Thermal conductivity

1 Btu/ft h deg R = 1.731 J/m s ^{o}C = 1.731 W/(mK)

1 cal/cm s deg K = 418.7 J/m s ^{o}C = 418.7 W/(mK)

Force

1 pdl = 0.1383 N

1 lbf = 32.17 pdl = 4.448 N

1 tonf = 9964 N

1 kgf = 2.205 lbf = 9.807 N

1 dyne = 10^{-5} N

Torque

1 lbf ft = 1.356 Nm

1 tonf ft = 3037 Nm

Power

1 hp = 550 ft lbf/s = 0.7457 kW

1 ft lbf/s = 1.356 W

1 metric horsepower (ch, PS) = 0.7355 kW

Energy, work, heat

1 ft lbf = 1.356 J

1 kW h = 3.6 MJ

1 Btu = 778.2 ft lbf = 252 cal = 1055 J

1 cal = 4.187 J

1 hp h = 2.685 MJ

Pressure, stress

1 lbf/in^2 = 0.07031 kgf/cm^2 = 6895 N/m^2

1 tonf/in^2 = 157.5 kgf/cm^2 = 15.44 MN/m^2

1 kgf/cm^2 = 0.09807 MN/m^2 = 0.9807 bar

1 kgf/mm^2 = 9.807 MN/m^2 = 0.9807 hbar

1 lbf/ft^2 = 47.88 N/m^2

1 ft H_2O = 62.43 lbf/ft^2 = 2989 N/m^2

1 in Hg = 70.73 lbf/ft^2 = 3386 N/m^2

1 mm Hg = 1 torr = 133.3 N/m^2

1 bar = 14.50 lbf/in^2 = 10^5 N/m^2

1 Int atm = 14.70 lbf/in^2 = 10.34 m water = 1.013 x 10^5 N/m^2
= 1.013 bar = 760 mm Hg = 101.3 kPa

Dynamic viscosity

1 poise (g/cm s)	=	0.1 kg/m s = 0.1 N s/m^2 = 0.1 Pa s
1 kgf s/m^2	=	0.9807 N s/m^2
1 lb/ft h	=	0.4132 mN s/m^2
1 slug/ft s	=	1 lbf s/ft^2 = 47.88 N s/m^2
1 lbf s/in^2	=	6895 N s/m^2

Kinematic viscosity

1 ft^2/s	=	0.09290 m^2/s
1 in^2/s	=	645.2 mm^2/s
1 cSt	=	1 mm^2/s

Electrical units

The conversion factors which follow are from the C.G.S. system to the SI system. (Note: in the C.G.S. system 1 e.m.u. = 3 x 10^{10} e.s.u. of charge).

capacitance	1 e.s.u.	= $\frac{1}{9}$ x 10^{-11} F
charge	1 e.m.u.	= 10 C
current	1 e.m.u.	= 10 A
electric field strength	1 e.s.u.	= 3 x 10^4 V/m
electric flux density	1 e.s.u.	= $\frac{1}{12\pi}$ x 10^{-5} C/m^2
electric polarisation	1 e.s.u.	= $\frac{1}{3}$ x 10^{-5} C/m^2
inductance	1 e.m.u.	= 10^{-9} H
intensity of magnetisation	1 e.m.u.	= 10^3 A/m
magnetic field strength	1 e.m.u.	= $\frac{1}{4\pi}$ x 10^3 A/m
magnetic flux	1 e.m.u.	= 10^{-8} Wb
magnetic flux density	1 e.m.u.	= 10^{-4} Wb/m^2
magnetic moment	1 e.m.u.	= 10^{-3} A m^2
magnetomotive force	1 e.m.u.	= $\frac{10}{4\pi}$ A
mass susceptibility	1 e.m.u/g	= 4π x 10^{-3} kg^{-1}
potential	1 e.m.u.	= 10^{-8} V
resistance	1 e.m.u.	= 10^{-9} Ω

2. PHYSICAL CONSTANTS

Avogadro's number	N	$= 6.023 \times 10^{26}/(\text{kg mol})$
Bohr magneton	β	$= 9.27 \times 10^{-24}$ A m^2
Boltzmann's constant	k	$= 1.380 \times 10^{-23}$ J/K
Stefan-Boltzmann constant	σ	$= 5.67 \times 10^{-8}$ W/(m^{2}K^4)
characteristic impedance of free space	Z_o	$= (\mu_o/\varepsilon_o)^{\frac{1}{2}} = 120\pi\Omega$
electron volt	eV	$= 1.602 \times 10^{-19}$ J
electron charge	e	$= 1.602 \times 10^{-19}$ C
electronic rest mass	m_e	$= 9.109 \times 10^{-31}$ kg
electronic charge to mass ratio	e/m_e	$= 1.759 \times 10^{11}$ C/kg
Faraday constant	F	$= 9.65 \times 10^{7}$ C/(kg mol)
permeability of free space	μ_o	$= 4\pi \times 10^{-7}$ H/m
permittivity of free space	ε_o	$= 8.85 \times 10^{-12}$ F/m
Planck's constant	h	$= 6.626 \times 10^{-34}$ J s
proton mass	m_p	$= 1.672 \times 10^{-27}$ kg
proton to electron mass ratio	m_p/m_e	$= 1836.1$
standard gravitational acceleration	g	$= 9.80665$ m/s^2 $= 9.80665$ N/kg
universal constant of gravitation	G	$= 6.67 \times 10^{-11}$ N m^2/kg^2
universal gas constant	R_o	$= 8.314$ kJ/(kg mol K)
velocity of light in vacuo	c	$= 2.9979 \times 10^{8}$ m/s
volume of 1 kg mol of ideal gas at 1 atm, 0°C		$= 22.41$ m^3

Temperature

$$^oC = \frac{5}{9} (^oF - 32)$$

$$K = \frac{5}{9} (^oF + 459.67) = \frac{5}{9} \; ^oR = \; ^oC + 273.15$$

3 . SUMMARY OF "BASIC"

This language contains the facilities provided in most versions of extended BASIC. Some instructions may vary somewhat from one system to another; however, equivalents should be available. This applies particularly to String Functions, Commands and Control Codes, and to items marked with a †. We suggest that you modify the SYSTEM DEPENDENT INSTRUCTIONS MARKED BY † to conform to your own system and add other instructions in the spaces provided.

Arithmetic Variable Names

numeric variables: e.g. A,X;B4,Z1

arithmetic array variables: e.g. S(4),A(I+1),N2(1,J), C(1,B(1))

String Variable Names

character string variables: e.g. B$

character string array variables: e.g. Z$(4),N$(A,B)

N.B. $ may be £ on some terminals. Use the key 'shift 4'.

Arithmetic Operators

↑ exponentiation e.g. 2↑3 gives 8

\- unary minus

* / multiplication, division

\+ - addition, subtraction

Operations inside any given pair of brackets are performed before those outside. Subject to this, BASIC performs operations in the order of the operators above. The only(†) exception is A↑-B, interpreted as A↑(-B). Operators of equal priority are applied from left to right.

e.g. 2↑(1+3/2*(1+1)) gives 16

Relational Operators (operate upon arithmetic and string values)

=	>
<	>=
<= (less than equal to)	<> (not equal to)

Logical Operators

AND

XOR exclusive

OR inclusive

Matrix Operators

+ -	addition or subtraction of matrices of equal dimensions
*	multiplication of conformable matrices
*	multiplication of a matrix by a scalar e.g. MAT A = (K)*A

Arithmetic Functions (x represents any expression)

PI	has the constant value 3.1415927
SIN(x),COS(x),TAN(x)	sine, cosine, tangent (x in radians)
ATN(x)	arctan (radians)
LOG(x),LOG10(x)	natural log, common log
EXP(x)	exponentiation e↑x where e = 2.71828
SQR(x)	square root
SGN(x)	sign of x (+ve gives 1, 0 gives 0, -ve gives -1)
ABS(x)	absolute value of x (|x|)
INT(x)	largest integer < = x
RND or RND(x)	returns a random number between 0 and 1. x, if present, is ignored.

† String Functions

LEN(A$)	returns the number of characters in the string A$, including trailing blanks
SUB$(A$,N1,N2)	creates a sub string from the string A$ starting with the N1th character and N2 characters long
SUB$(A$,N)	creates a sub string from the string A$, starting with the Nth character to the last character in A$
*CHR$(x)	returns a one character string having the ASCII value x
*ASCII(A$)	returns the ASCII value of the first character in A$
NUM$(N)	creates the string of characters that would be printed by PRINT N;
NUM$(N,field)	creates the string of characters that would be printed by PRINTUSING"field",N;
VAL(A$)	computes the value that would be generated by the INPUT of the characters of A$ to an arithmetic variable

*The ASCII value is based on seven bit characters. Treatment of the parity bit is system dependent.

†Error functions (only valid in an error handling routine entered by ONERROR)

ERR	contains the error number of the most recent error
ERL	contains the line number of the most recent error

Matrix functions

MAT Y = TRN(X)	Y becomes the transpose of X
MAT Y = INV(X)	Y becomes the inverse of X
DET	contains the determinant of X after the evaluation of INV(X)

User defined functions - see DEF statement

Statements

Note - a program line may contain several statements separated by the colon (:) character

Type	Example
CLOSE	CLOSE 2
DATA	DATA 4.3,85,"MONDAY"
DEF	DEF FNA(X) = X↑X DEF FNA1(A,B) = SQR(A+B) † DEF FNF(M) IF M = 1 THEN FNF = 1 ELSE FNF = M*FNF(M-1) FNEND
DIM	DIM A(10),B$(5,10)
END	END must be the last statement of a program
FOR	FOR X = 1 TO 10 FOR N = A TO A+R FOR 1 = 2 to 40 STEP 2
GOSUB	GOSUB 200
GOTO	GOTO 151
IF	IF B = A THEN 21 IF A > Z THEN PRINT "BIGGER" IF R < N+1 THEN R = N ELSE R = N+2 IF A > B OR B < C THEN STOP IF FNA(R) = B GOTO 200
INPUT	INPUT A INPUT "TYPE YOUR NAME",A$ INPUT # 4,N,M
LET	LET A = 20 LET A,B,C = 0 A$ = "TEXT" (LET is optional)

MAT	MAT C = CON all elements of C = 1 MAT B = IDN(10,10) identity matrix MAT A = ZER all elements of A = 0 MAT B = ZER(5,10) redimensions and zeros B
MATINPUT	MATINPUT A,B,C(4) MATINPUT # 3,A,C
MATPRINT	MATPRINT B MATPRINT B(10,5); MATPRINT # 2,A
MATREAD	MATREAD A,B(4,4)
NEXT	NEXT I
ON ERROR GOTO	ON ERROR GOTO 140
ON GOSUB	ON X GOSUB 200,250,300 ON FNA(A) + FNB(A) GOSUB 10,15,30,5
ON GOTO	ON A + 1 GOTO 14,25,50
OPEN	†
PRINT	PRINT A,B PRINT "RESULT": X1 PRINT # 4, I*A,."EXPERIMENT"; N
† PRINT USING	PRINT USING "# # #", A,B PRINT # 3, USING B$,C,Z$ PRINT USING 1000,X
RANDOM	RANDOM
READ	READ A,B$,F1,C
REM	REMARK THIS IS A COMMENT
!	An exclamation mark at the beginning of a line is equivalent to REM An exclamation mark after any statement causes the rest of the line to be treated as comment X = 0! ZERO CONTROL
RESTORE	RESTORE
RESUME	RESUME RESUME 240
RETURN	RETURN
STOP	STOP
TRACE	TRACE
† %	PRINTUSING image, e.g. 1000% X = # #.#

†Commands

RUN	runs the current program
LOAD	loads a program from paper tape (or other medium)
CLEAR NEW	remove any existing program
LIST	LIST prints the current program LIST n prints line n LIST n-m prints lines n to m
DELETE	DELETE 40-45 deletes specified lines from the current program
SAVE	saves a program on paper tape (or other medium)
REP	edits a program line. Any non-numerical character can be used as separator e.g. REP10/X1/A REP30/,B/
RESEQUENCE	renumbers part or all of a program (including GOTO etc references) e.g. RESEQUENCE whole program, steps of 10 RESEQUENCE 900,1000 after old line 900, which becomes 1000, steps of 10 RESEQUENCE,,5 whole program, steps of 5

†Special control codes

ESC ACCEPT	breaks into the program and stops it, typing BASIC READY
CR RETURN	terminate a line of input
\ or $	(on the same key as L, not 4) Abandon the current line of input
← or RUBOUT	delete the previous character or space (may be used repeatedly)

4. ANALYSIS

4.1 Vector algebra

$$\hat{\underline{a}} = \underline{a}/|\underline{a}|$$

$$\underline{a} = a_1\underline{i} + a_2\underline{j} + a_3\underline{k} \equiv (a_1, a_2, a_3)$$

$$a \equiv |\underline{a}| = \sqrt{a_1^2 + a_2^2 + a_3^2}$$

$$\underline{a} + \underline{b} = (a_1+b_1, a_2+b_2, a_3+b_3)$$

Scalar (dot) product:

$$\underline{a} \cdot \underline{b} = a_1b_1 + a_2b_2 + a_3b_3 = ab\mathrm{Cos}\theta$$

Vector (cross) product:

$$\underline{a} \times \underline{b} = \begin{vmatrix} \underline{i} & \underline{j} & \underline{k} \\ a_1 & a_2 & a_3 \\ b_1 & b_2 & b_3 \end{vmatrix} = ab\mathrm{Sin}\theta\hat{\underline{n}}$$

where $\hat{\underline{n}} \perp \underline{a}$, $\hat{\underline{n}} \perp \underline{b}$

Triple scalar product:

$$[\underline{a}\ \underline{b}\ \underline{c}] = \underline{a} \cdot \underline{b} \times \underline{c} = \underline{a} \times \underline{b} \cdot \underline{c} = \begin{vmatrix} a_1 & a_2 & a_3 \\ b_1 & b_2 & b_3 \\ c_1 & c_2 & c_3 \end{vmatrix}$$

Triple vector product:

$$\underline{a} \times (\underline{b} \times \underline{c}) = (\underline{a}.\underline{c})\underline{b} - (\underline{a}.\underline{b})\underline{c}$$

$$(\underline{a} \times \underline{b}) \times \underline{c} = (\underline{a}.\underline{c})\underline{b} - (\underline{b}.\underline{c})\underline{a}$$

Differentiation of vectors:

$$\frac{d}{dt}(\underline{a} + \underline{b}) = \frac{d\underline{a}}{dt} + \frac{d\underline{b}}{dt} \qquad \frac{d}{dt}(f\ \underline{a}) = \frac{df}{dt}\underline{a} + f\frac{d\underline{a}}{dt}$$

$$\frac{d}{dt}(\underline{a}.\underline{b}) = \underline{a}.\frac{d\underline{b}}{dt} + \frac{d\underline{a}}{dt}.\underline{b} \qquad \frac{d}{dt}(\underline{a} \times \underline{b}) = \underline{a} \times \frac{d\underline{b}}{dt} + \frac{d\underline{a}}{dt} \times \underline{b}$$

$$\frac{d}{dt}(\underline{a}.\underline{b} \times \underline{c}) = \frac{d\underline{a}}{dt}.\underline{b} \times \underline{c} + a.\frac{d\underline{b}}{dt} \times \underline{c} + \underline{a}.\underline{b} \times \frac{d\underline{c}}{dt}$$

Gradient:

$$\text{grad } V = \nabla V = \underline{i}\frac{\partial V}{\partial x} + \underline{j}\frac{\partial V}{\partial y} + \underline{k}\frac{\partial V}{\partial z} \quad \text{(Cartesian)}$$

$$= \underline{u}_r\frac{\partial V}{\partial r} + \underline{u}_\phi \frac{1}{r}\frac{\partial V}{\partial \phi} + \underline{u}_z\frac{\partial V}{\partial z} \quad \text{(Cylindrical)}$$

where
$$\underline{u}_r = \underline{i}\cos\phi + \underline{j}\sin\phi$$
$$\underline{u}_\phi = -\underline{i}\sin\phi + \underline{j}\cos\phi$$
$$\underline{u}_z = \underline{k}$$

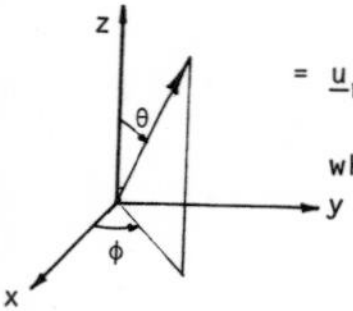

$$= \underline{u}_r\frac{\partial V}{\partial r} + \frac{\underline{u}_\theta}{r}\frac{\partial V}{\partial \theta} + \frac{\underline{u}_\phi}{r\sin\theta}\frac{\partial V}{\partial \phi} \quad \text{(Spherical)}$$

where
$$\underline{u}_r = \underline{i}\cos\phi\sin\theta + \underline{j}\sin\phi\sin\theta + \underline{k}\cos\theta$$
$$\underline{u}_\theta = \underline{i}\cos\phi\cos\theta + \underline{j}\sin\phi\cos\theta - \underline{k}\sin\theta$$
$$\underline{u}_\phi = -\underline{i}\sin\phi + \underline{j}\cos\phi$$

Divergence:

$$\text{div } \underline{F} = \nabla.\underline{F} = \frac{\partial F_x}{\partial x} + \frac{\partial F_y}{\partial y} + \frac{\partial F_z}{\partial z} \quad \text{(Cartesian)}$$

$$= \frac{1}{r}\frac{\partial}{\partial r}\left(rF_r\right) + \frac{1}{r}\frac{\partial F_\phi}{\partial \phi} + \frac{\partial F_z}{\partial z} \quad \text{(Cylindrical)}$$

$$= \frac{1}{r^2}\frac{\partial}{\partial r}\left(r^2F_r\right) + \frac{1}{r\sin\theta}\frac{\partial}{\partial \theta}(F_\theta\sin\theta) + \frac{1}{r\sin\theta}\frac{\partial F_\phi}{\partial \phi} \quad \text{(Spherical)}$$

Curl:

$$\text{curl } \underline{F} = \nabla\times\underline{F} = \underline{i}\left(\frac{\partial F_z}{\partial y} - \frac{\partial F_y}{\partial z}\right) + \underline{j}\left(\frac{\partial F_x}{\partial z} - \frac{\partial F_z}{\partial x}\right) + \underline{k}\left(\frac{\partial F_y}{\partial x} - \frac{\partial F_x}{\partial y}\right) \quad \text{(Cartesian)}$$

$$= \begin{vmatrix} \underline{i} & \underline{j} & \underline{k} \\ \frac{\partial}{\partial x} & \frac{\partial}{\partial y} & \frac{\partial}{\partial z} \\ F_x & F_y & F_z \end{vmatrix}$$

$$= \frac{1}{r}\begin{vmatrix} \underline{u}_r & r\underline{u}_\phi & \underline{u}_z \\ \frac{\partial}{\partial r} & \frac{\partial}{\partial \phi} & \frac{\partial}{\partial z} \\ F_r & rF_\phi & F_z \end{vmatrix} \quad \text{(Cylindrical)}$$

$$= \frac{1}{r^2\sin\theta}\begin{vmatrix} \underline{u}_r & r\underline{u}_\theta & r\sin\theta\,\underline{u}_\phi \\ \frac{\partial}{\partial r} & \frac{\partial}{\partial \theta} & \frac{\partial}{\partial \phi} \\ F_r & rF_\theta & r\sin\theta F_\phi \end{vmatrix} \quad \text{(Spherical)}$$

Laplace:

$$\nabla.\nabla V = \nabla^2 V = \frac{\partial^2 V}{\partial x^2} + \frac{\partial^2 V}{\partial y^2} + \frac{\partial^2 V}{\partial z^2} \quad \text{(Cartesian)}$$

$$= \frac{1}{r}\frac{\partial}{\partial r}\left(r\frac{\partial V}{\partial r}\right) + \frac{1}{r^2}\frac{\partial^2 V}{\partial \phi^2} + \frac{\partial^2 V}{\partial z^2} \quad \text{(Cylindrical)}$$

$$= \frac{1}{r^2}\frac{\partial}{\partial r}\left(r^2\frac{\partial V}{\partial r}\right) + \frac{1}{r^2\sin\theta}\frac{\partial}{\partial \theta}\left(\sin\theta\frac{\partial V}{\partial \theta}\right) + \frac{1}{r^2\sin^2\theta}\frac{\partial^2 V}{\partial \phi^2}$$

(Spherical)

Space curves:

$\underline{v} = \underline{u}\frac{ds}{dt}$, s = arc length $\underline{u}$ = unit tangent

$\underline{a} = \frac{v^2}{\rho}\underline{n} + \frac{dv}{dt}\underline{u}$ $\underline{n}$ = unit 'inward' normal

$\frac{d\underline{u}}{ds} = \frac{1}{\rho}\underline{n}$ ρ = radius of curvature

$\underline{b} = \underline{u} \times \underline{n}$, $\underline{b}$ = binormal vector

$\frac{d\underline{b}}{ds} = -\frac{1}{\tau}\underline{n}$, $\frac{d\underline{n}}{ds} = \frac{1}{\tau}\underline{b} - \frac{1}{\rho}\underline{u}$, $\frac{1}{\tau}$ = torsion

Identities:

$\nabla.\phi\underline{u} = \phi\nabla.\underline{u} + \underline{u}.\nabla\phi$

$\nabla\times\phi\underline{u} = \phi\underline{\nabla}\times\underline{u} + \underline{\nabla}\phi\times\underline{u}_o$

$\nabla.\underline{u}\times\underline{v} = \underline{v}.\nabla\times\underline{u} - \underline{u}.\nabla\times\underline{v}$

4.2 Series

$$(1+x)^\alpha = 1 + \alpha x + \frac{\alpha(\alpha-1)}{2!}x^2 + \frac{\alpha(\alpha-1)(\alpha-2)}{3!}x^3 + \ldots,$$

for arbitrary α, $|x| < 1$

$$e^x = 1 + x + \frac{x^2}{2!} + \ldots + \frac{x^n}{n!} + \ldots \quad \text{for all } x$$

$$\cos x = 1 - \frac{x^2}{2!} + \frac{x^4}{4!} - \dots + \frac{(-1)^n}{(2n)!} x^{2n} + \dots \text{ for all } x$$

$$\sin x = x - \frac{x^3}{3!} + \frac{x^5}{5!} - \dots + \frac{(-1)^n}{(2n+1)!} x^{2n+1} + \dots \text{ for all } x$$

$$\tan x = x + \frac{x^3}{3} + \frac{2x^5}{15} + \frac{17x^7}{315} + \dots \text{ for } |x| < \pi/2$$

$$\ell n(1+x) = x - \frac{x^2}{2} + \frac{x^3}{3} - \dots + \frac{(-1)^n}{(n+1)} x^{n+1} + \dots$$

$$\text{for } -1 < x \leq 1$$

Taylor's

$$f(a+h) = f(a) + hf'(a) + \frac{h^2}{2!} f''(a) + \dots$$

$$+ \frac{h^{n-1}}{(n-1)!} f^{(n-1)}(a) + \frac{h^n}{n!} f^{(n)}(c) \text{ where } a < c < a+h$$

Maclaurin's

$$f(x) = f(o) + xf'(o) + \frac{x^2}{2!} f''(o) + \dots$$

$$+ \frac{x^{n-1}}{(n-1)!} f^{(n-1)}(o) + \frac{x^n}{n!} f^{(n)}(\theta x) \text{ where } o < \theta < 1$$

Stirling's formula for n!

For n large, $n! \sim \sqrt{(2\pi)}\, n^{n+\frac{1}{2}} e^{-n}$

or, $\log_{10} n! \approx 0.39909 + (n+\frac{1}{2})\log_{10} n - 0.43429n.$

Fourier series

(i) General formulae

If $f(x)$ is periodic of period $2L$, $f(x+2L) = f(x)$

$$f(x) = \tfrac{1}{2}a_o + \sum_{n=1}^{\infty} a_n \cos\frac{n\pi x}{L} + \sum_{n=1}^{\infty} b_n \sin\frac{n\pi x}{L}$$

where

$$a_n = \frac{1}{L}\int_{-L}^{L} f(x)\cos\frac{n\pi x}{L}\,dx \quad n = 0, 1, 2, \dots$$

$$b_n = \frac{1}{L}\int_{-L}^{L} f(x) \sin \frac{n\pi x}{L}\, dx \qquad n = 1, 2, 3, \ldots$$

If $f(x)$ is an <u>even</u> function of x, i.e., $f(-x) = f(x)$

then $$a_n = \frac{2}{L}\int_{0}^{L} f(x) \cos \frac{n\pi x}{L}\, dx \qquad n = 0, 1, 2 \ldots$$

and $$b_n = 0 \qquad n = 1, 2, 3, \ldots$$

If $f(x)$ is an <u>odd</u> function of x, i.e., $f(-x) = -f(x)$

then $$a_n = 0 \qquad n = 0, 1, 2, \ldots$$

and $$b_n = \frac{2}{L}\int_{0}^{L} f(x) \sin \frac{n\pi x}{L}\, dx \qquad n = 1, 2, 3, \ldots$$

(ii) Special waveforms, all of period 2L

(a) Square wave, sine series

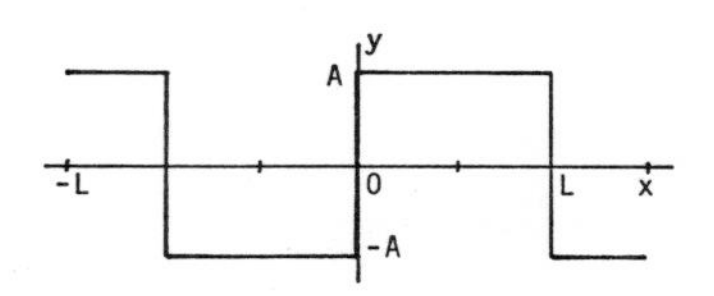

$$f(x) = \frac{4A}{\pi}\left[\sin \frac{\pi x}{L} + \frac{1}{3}\sin \frac{3\pi x}{L} + \frac{1}{5}\sin \frac{5\pi x}{L} + \ldots\right]$$

mean square value $= A^2$

(b) Square wave, cosine series

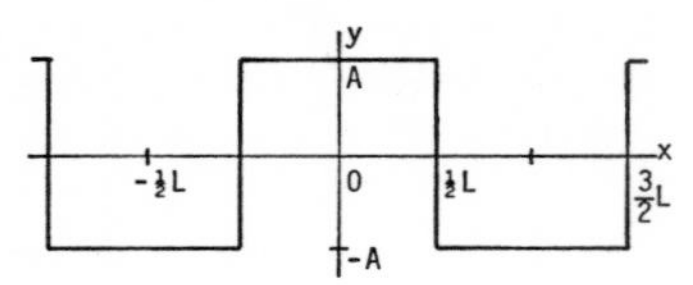

$$f(x) = \frac{4A}{\pi}\left[\cos\frac{\pi x}{L} - \frac{1}{3}\cos\frac{3\pi x}{L} + \frac{1}{5}\cos\frac{5\pi x}{L} - \ldots\right]$$

mean square value $= A^2$

(c) Triangular wave

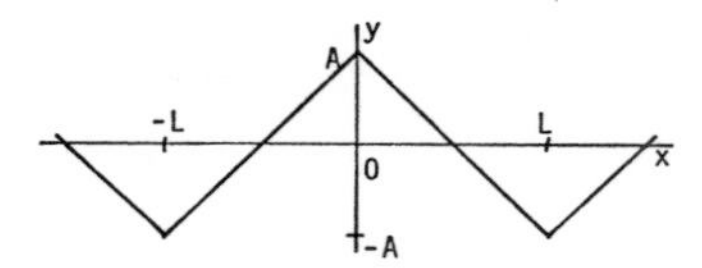

$$f(x) = \frac{8A}{\pi^2}\left[\cos\frac{\pi x}{L} + \frac{1}{3^2}\cos\frac{3\pi x}{L} + \frac{1}{5^2}\cos\frac{5\pi x}{L} + \ldots\right]$$

mean square value $= \frac{A^2}{3}$

(d) Saw-tooth wave

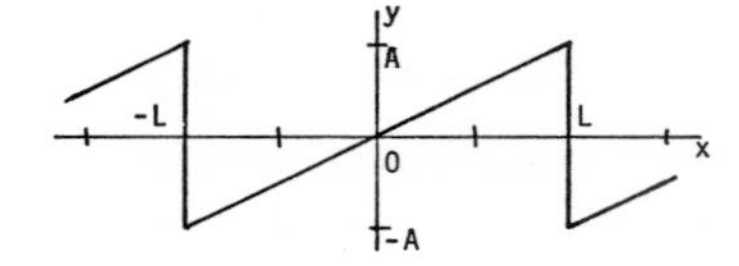

$$f(x) = \frac{2A}{\pi}\left[\sin\frac{\pi x}{L} - \frac{1}{2}\sin\frac{2\pi x}{L} + \frac{1}{3}\sin\frac{3\pi x}{L} - \ldots\right]$$

mean square value $= \frac{A^2}{3}$

(e) Half-wave rectification

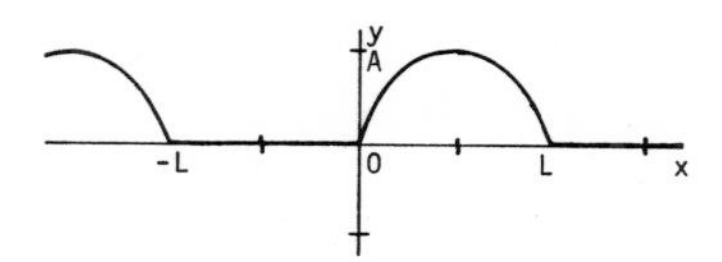

$$f(x) = \frac{A}{2} \sin \frac{\pi x}{L} + \frac{2A}{\pi} \left[\frac{1}{2} - \frac{1}{3} \cos \frac{2\pi x}{L} - \frac{1}{15} \cos \frac{4\pi x}{L} - \ldots\right]$$

mean square value $= \frac{A^2}{4}$ average value $= \frac{A}{\pi}$

(f) Full-wave rectification

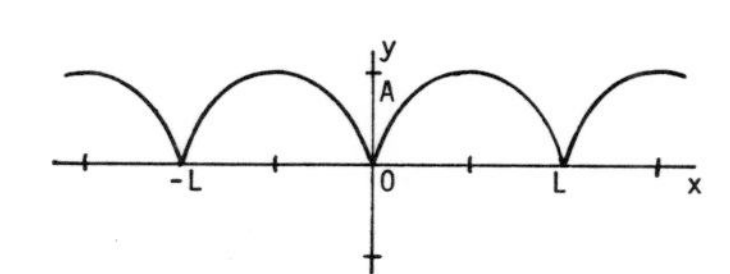

$$f(x) = \frac{4A}{\pi} \left[\frac{1}{2} - \frac{1}{3} \cos \frac{2\pi x}{L} - \frac{1}{15} \cos \frac{4\pi x}{L} - \ldots\right]$$

mean square value $= \frac{A^2}{2}$ average value $= \frac{2A}{\pi}$

4.3 Trigonometric, hyperbolic and algebraic relations

$$2 \sin A \cos B = \sin(A-B) + \sin(A+B)$$

$$2 \cos A \cos B = \cos(A-B) + \cos(A+B)$$

$$2 \sin A \sin B = \cos(A-B) - \cos(A+B)$$

$$\sin A + \sin B = 2 \sin\tfrac{1}{2}(A+B) \cos\tfrac{1}{2}(A-B)$$

$$\sin A - \sin B = 2 \cos\tfrac{1}{2}(A+B) \sin\tfrac{1}{2}(A-B)$$

$$\cos A + \cos B = 2 \cos\tfrac{1}{2}(A+B) \cos\tfrac{1}{2}(A-B)$$

$$\cos A - \cos B = -2 \sin\tfrac{1}{2}(A+B) \sin\tfrac{1}{2}(A-B)$$

$$\sin(A \pm B) = \sin A \cos B \pm \cos A \sin B$$

$$\cos(A \pm B) = \cos A \cos B \mp \sin A \sin B$$

$$\tan(A \pm B) = \frac{\tan A \pm \tan B}{1 \mp \tan A \tan B}$$

$$\sin^2 A + \cos^2 A = 1$$

$$\mathrm{Sec}^2 A = \mathrm{Tan}^2 A + 1$$

$$\frac{a}{\sin A} = \frac{b}{\sin B} = \frac{c}{\sin C}$$

$$\sin A = \frac{2}{bc}\sqrt{s(s-a)(s-b)(s-c)} \text{ where } s = \tfrac{1}{2}(a+b+c)$$

$$= \frac{2}{bc}\text{ area}$$

$$a^2 = b^2 + c^2 - 2\,bc\cos A$$

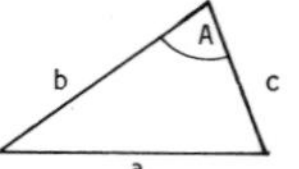

Relation for Spherical Triangles

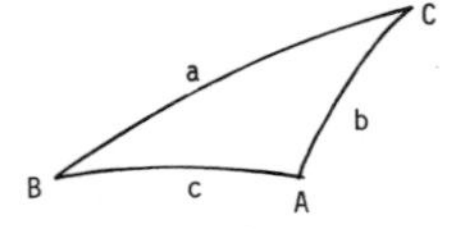

$$\frac{\sin a}{\sin A} = \frac{\sin b}{\sin B} = \frac{\sin c}{\sin C}$$

$$\cos a = \cos b \cos c + \sin b \sin c \cos A$$

$$\cos A = -\cos B \cos C + \sin B \sin C \cos a$$

$$\sin\frac{A}{2} = \sqrt{\frac{\sin(s-b)\sin(s-c)}{\sin b \sin c}} \quad \text{where } s = \tfrac{1}{2}(a+b+c)$$

$$\sin\frac{a}{2} = \sqrt{-\frac{\cos S \cos(S-A)}{\sin B \sin C}} \quad \text{where } S = \tfrac{1}{2}(A+B+C)$$

Napiers Rules for right spherical triangles:

Arrange the five parts about the right angle with 'co' attached to the three parts opposite the right angle. E.g. for the right angle at A we have

co-a | co-C | b | c | co-B — Right angle

N.B. co-a is the complement of a. i.e. 90^0-a

Then: The sine of the middle part is the product of the tangents of adjacent parts and is the product of the cosines of opposite parts.

N.B. A leg and its opposite angle are always in the same quadrant. If the hypotenuse is less than 90^0 the legs are in the same quadrant, otherwise they are in opposite quadrants.

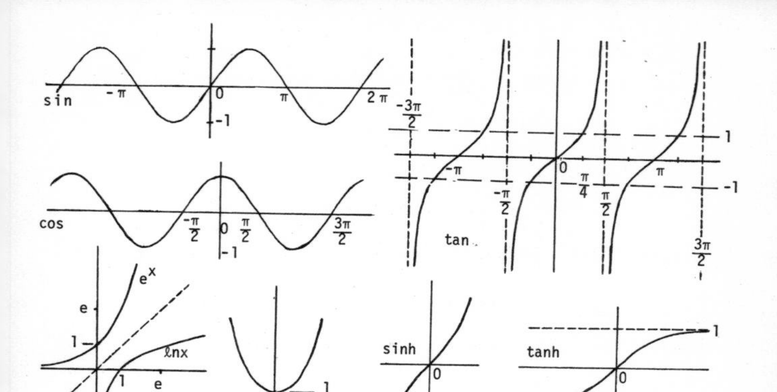

$$\sin x = \frac{e^{ix} - e^{-ix}}{2i} \qquad \cos x = \frac{e^{ix} + e^{-ix}}{2}$$

$$\sinh x = \frac{e^{x} - e^{-x}}{2} \qquad \cosh x = \frac{e^{x} + e^{-x}}{2}$$

$$\cos iz = \cosh z \qquad \sin iz = i \sinh z$$

$$\cosh iz = \cos z \qquad \sinh iz = i \sin z$$

$$e^z = \cosh z + \sinh z$$

$$\log_{10}(10^x) \equiv \log_{10}(\text{antilog}_{10} x) \equiv x \equiv 10^{\log_{10} x} \equiv e^{\log_e x} = e^{\ell n x}$$

$$a^2 - b^2 = (a+b)(a-b) \; : \; a^3 - b^3 = (a-b)(a^2+ab+b^2)$$

$$x = \frac{-b \pm \sqrt{(b^2 - 4ac)}}{2a}$$

equations of curves

circle	ellipse	hyperbola	parabola
$x^2 + y^2 = a^2$	$\frac{x^2}{a^2} + \frac{y^2}{b^2} = 1$	$\frac{x^2}{a^2} - \frac{y^2}{b^2} = 1$	$y^2 = ax$

4.4 Complex numbers

$$z = r(\cos\theta + i\sin\theta) = x + iy$$

$$= r\,e^{i(\theta+2n\pi)} \qquad (n = 0, \pm1, \pm2, \ldots.)$$

$$e^{iz} = \cos z + i\sin z \quad \text{[Euler's Formula]}$$

$$x + iy = \sqrt{x^2 + y^2}\; e^{i\tan^{-1}(y/x)}; \quad z^c = e^{c\ln z}$$

N.B. $\tan^{-1}(y/x)$ must be chosen to lie in the appropriate quadrant

4.5 Partial differentiation

(a) If $F = f(x,y)$, where $x = X(t)$, $y = Y(t)$ then

$$F = F(t) \text{ and } \frac{dF}{dt} = \frac{\partial f}{\partial x}\frac{dX}{dt} + \frac{\partial f}{\partial y}\frac{dY}{dt}$$

(b) If $F = f(x,y)$, where $y = Y(x)$, then $F = F(x)$ and

$$\frac{dF}{dx} = \frac{\partial f}{\partial x} + \frac{\partial f}{\partial y}\frac{dY}{dx}$$

(c) If $F = f(x,y)$, where $x = X(u,v)$, $y = Y(u,v)$ then

$$F = F(u,v) \text{ and } \frac{\partial F}{\partial u} = \frac{\partial f}{\partial x}\frac{\partial X}{\partial u} + \frac{\partial f}{\partial y}\frac{\partial Y}{\partial u},$$

$$\frac{\partial F}{\partial v} = \frac{\partial f}{\partial x}\frac{\partial X}{\partial v} + \frac{\partial f}{\partial y}\frac{\partial Y}{\partial v}.$$

4.6 Differential Equations

(i) First Order

Type	Characteristic	Method of solution
separable	$y' = P(x)Q(y)$	rearrange:- $\int\frac{1}{Q}\,dy = \int P\,dx + c$
homogeneous	$y' = f\left(\frac{y}{x}\right)$	by substitution $y = ux$ to make equation separable
exact	$M(x,y)dx + N(x,y)dy$ where $\frac{\partial M}{\partial y} = \frac{\partial N}{\partial x}$	$\frac{\partial G}{\partial x} = M$, $\frac{\partial G}{\partial y} = N$ Solve for G
linear	$y' + P(x)y = Q(x)$	multiply through by $e^{\int P dx}$

(ii) <u>Second Order</u>, linear with constant coefficients

$$m\ddot{x} + \alpha\dot{x} + kx = 0$$

$$\ddot{x} + 2\xi\omega_o\dot{x} + \omega_o^2 x = 0, \quad \xi = \frac{\alpha}{2\sqrt{mk}}, \quad \omega_o = \sqrt{\frac{k}{m}}$$

(a) $\xi < 1$ (underdamping)

$$x = a\, e^{-\xi\omega_o t}\cos(\omega t - \theta)$$

$$\omega = \omega_o\sqrt{1-\xi^2}$$

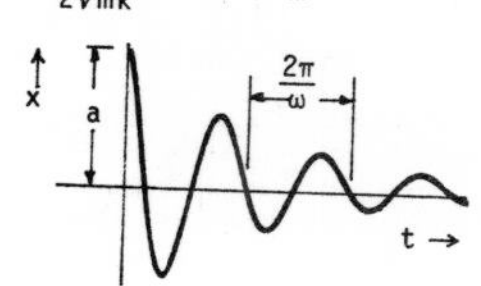

(b) $\xi > 1$ (overdamping)

$$x = A_1 e^{-q_1 t} + A_2 e^{-q_2 t}$$

where $q_1, q_2 = \omega_o\{\xi \pm \sqrt{(\xi^2 - 1)}\}$

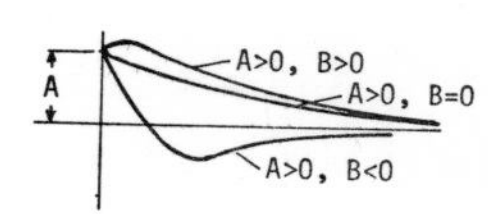

(c) $\xi = 1$ (critical damping)

$$x = (A + Bt)e^{-\xi\omega_o t}$$

A
A>0, B>0
A>0, B=0
A>0, B<0

<u>Forced oscillations</u>

$$\ddot{x} + 2\xi\omega_o\dot{x} + \omega_o^2 x = a\cos pt, \quad a = \frac{F}{m}, \quad x_1 = \frac{F}{k}$$

$$x = \frac{Aa\cos(pt - \phi)}{\omega_o^2}$$

$$\tan\phi = \frac{2\xi\, p/\omega_o}{1 - \left(\frac{p}{\omega_o}\right)^2}$$

$$A = \left|\frac{x}{x_1}\right| = \frac{1}{\left[\left[1 - \left(\frac{p}{\omega_o}\right)^2\right]^2 + \left(2\xi\frac{p}{\omega_o}\right)^2\right]^{\frac{1}{2}}}$$

At resonance $p = \omega_o\sqrt{1 - 2\xi^2}$

$$x = \frac{x_1}{2\xi\sqrt{1-\xi^2}} \qquad \tan\phi = \frac{\sqrt{1-2\xi^2}}{\xi}$$

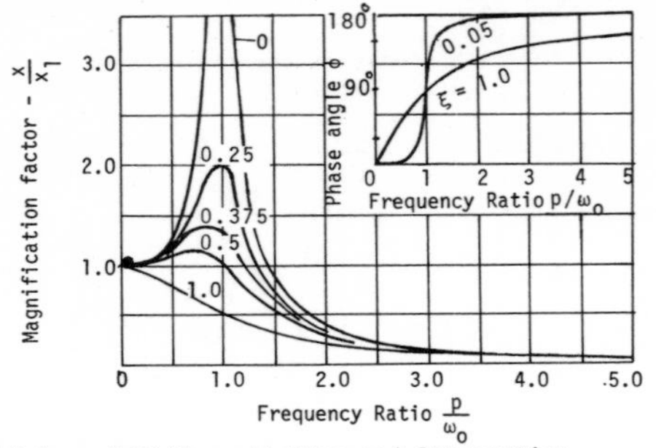

4.7 Rules of Differentiation and Integration

$$\frac{d}{dx}(uv) = u\frac{dv}{dx} + v\frac{du}{dx}$$

$$\frac{d}{dx}(uvw) = uv\frac{dw}{dx} + uw\frac{dv}{dx} + vw\frac{du}{dx}$$

$$\frac{d}{dx}\left(\frac{u}{v}\right) = \frac{1}{v^2}\left(v\frac{du}{dx} - u\frac{dv}{dx}\right)$$

$$\int uv\,dx = uw - \int\frac{du}{dx}w\,dx, \text{ where } w = \int v\,dx$$

4.8 Standard Differentials and Integrals

$$\frac{d}{dx}x^n = nx^{n-1} \qquad \int x^n\,dx = \frac{x^{n+1}}{n+1}, \quad n \neq -1$$

$$\frac{d}{dx}\ell n|x| = \frac{1}{x} \qquad \int\frac{dx}{x} = \ell n|x|$$

$$\frac{d}{dx}e^{ax} = ae^{ax} \qquad \int e^{ax}\,dx = \frac{e^x}{a} \qquad a \neq 0$$

$$\frac{d}{dx}a^x = a^x\,\ell n\,a \qquad \int a^x\,dx = \frac{a^x}{\ell n\,a}, \quad a > 0, \quad a \neq 1$$

$$\frac{d}{dx}x^x = x^x(1+\ell n\,x)$$

$$\int\ell n\,x\,dx = x(\ell n\,x - 1)$$

$$\frac{d}{dx}\sin x = \cos x \qquad \int \cos x\, dx = \sin x$$

$$\frac{d}{dx}\cos x = -\sin x \qquad \int \sin x\, dx = -\cos x$$

$$\frac{d}{dx}\tan x = \sec^2 x \qquad \int \sec^2 x\, dx = \tan x$$

$$\frac{d}{dx}\cot x = -\mathrm{cosec}^2 x \qquad \int \mathrm{cosec}^2 x\, dx = -\cot x$$

$$\frac{d}{dx}\sin^{-1} x = \frac{1}{\sqrt{(1-x^2)}} \qquad \int \frac{dx}{\sqrt{(1-x^2)}} = \sin^{-1} x, \quad |x| < 1$$

$$\frac{d}{dx}\tan^{-1} x = \frac{1}{1+x^2} \qquad \int \frac{dx}{1+x^2} = \tan^{-1} x$$

$$\frac{d}{dx}\cosh x = \sinh x \qquad \int \sinh x\, dx = \cosh x$$

$$\frac{d}{dx}\sinh x = \cosh x \qquad \int \cosh x\, dx = \sinh x$$

$$\frac{d}{dx}\tanh x = \mathrm{sech}^2 x \qquad \int \mathrm{sech}^2 x\, dx = \tanh x$$

$$\frac{d}{dx}\coth x = -\mathrm{cosech}^2 x \qquad \int \mathrm{cosech}^2 x\, dx = -\coth x$$

$$\frac{d}{dx}\sinh^{-1} x = \frac{1}{\sqrt{(1+x^2)}} \qquad \int \frac{dx}{\sqrt{(1+x^2)}} = \sinh^{-1} x$$

$$= \ln |x + \sqrt{(1+x^2)}|$$

$$\frac{d}{dx}\cosh^{-1} x = \frac{1}{\sqrt{(x^2-1)}} \qquad \int \frac{dx}{\sqrt{(x^2-1)}} = \cosh^{-1} x$$

$$= \ln |x + \sqrt{(x^2-1)}|, x \geqslant 1$$

$$\frac{d}{dx}\tanh^{-1} x = \frac{1}{1-x^2} \qquad \int \frac{dx}{1-x^2} = \tanh^{-1} x$$

$$= \tfrac{1}{2} \ln \left|\frac{1+x}{1-x}\right|, \; x^2 < 1$$

$$\frac{d}{dx}\coth^{-1} x = \frac{1}{1-x^2} \qquad \int \frac{dx}{x^2-1} = -\coth^{-1} x$$

$$= \tfrac{1}{2} \ln \left|\frac{x-1}{x+1}\right|, \; x^2 > 1$$

Some definite integrals (m,n integers)

$$\int_0^{\frac{\pi}{2}} \sin^n x dx = \int_0^{\frac{\pi}{2}} \cos^n x dx = \begin{cases} \frac{n-1}{n} \; \frac{n-3}{n-2} \cdots \frac{3}{4} \; \frac{1}{2} \; \frac{\pi}{2}, & n \text{ even} \\ \frac{n-1}{n} \; \frac{n-3}{n-2} \cdots \frac{4}{5} \; \frac{2}{3} \; 1, & n \text{ odd} \end{cases}$$

$$I_{m,n} = \int_0^{\frac{\pi}{2}} \sin^m x \cos^n x dx = \left(\frac{m-1}{m+n}\right) I_{m-2,n} = \left(\frac{n-1}{m+n}\right) I_{m,n-2}, \; m \neq -n$$

$$\int_0^{\pi} \sin mx \sin nx \, dx = \int_0^{\pi} \cos mx \cos nx \, dx = 0 \quad (m \neq n)$$

$$\int_0^{\pi} \sin nx \cos nx \, dx = 0$$

$$\int_0^{\infty} e^{-ax} \sin bx dx = \frac{b}{a^2+b^2}, \; a > 0$$

$$\int_0^{\infty} e^{-ax} \cos bx dx = \frac{a}{a^2+b^2}, \; a > 0$$

$$\int_0^{\infty} e^{-x^2} dx = \frac{\sqrt{\pi}}{2}$$

The error function $\quad \operatorname{erf} z = \frac{2}{\sqrt{\pi}} \int_0^z e^{-u^2} du$ (refer to page 30 for tabulated value)

$$\int_0^{\pi} \frac{\sin\theta \cos\theta}{(1+\varepsilon\cos\theta)^3} d\theta = \frac{-2\varepsilon}{(1-\varepsilon^2)^2}$$

$$\int_0^{\pi} \frac{\sin^2 \theta d\theta}{(1+\varepsilon\cos\theta)^3} = \frac{\pi}{2(1-\varepsilon^2)^{3/2}}$$

$$\int_0^{2\pi} \frac{d\theta}{(1+\varepsilon\cos\theta)} = \frac{2\pi}{(1-\varepsilon^2)^{1/2}}$$

4.9 Laplace Transforms

Definition

$$F(s) = L[f(t)] = \int_0^\infty f(t)e^{-st}dt$$

Theorems

Linearity	$L[af(t)+bg(t)]$	$= aF(s)+bG(s)$
Final Value	$\lim_{t \to \infty} f(t)$	$= \lim_{s \to o} sF(s)$
Initial Value	$\lim_{t \to o} f(t)$	$= \lim_{s \to \infty} sF(s)$
Differentiation	$L\left[\frac{df(t)}{dt}\right]$	$= sF(s)-f(o)$
	$L\left[\frac{d^2f(t)}{dt^2}\right]$	$= s^2F(s)-sf(o)-f^1(o)$
Integration	$L[\int f(t)dt]$	$= \frac{F(s)}{s} + \frac{f^{-1}(o)}{s}$
First Shifting	$L[e^{at}f(t)]$	$= F(s-a)$
Second Shifting	$L[f(t-a)]$ $t>a$	$= e^{-as}F(s)$
Convolution $L[f*g] \equiv$	$L[\int_0^t f(u)g(t-u)du]$	$= F(s)G(s)$
Partial Differentiation	$L\left[\frac{\partial f(t,\alpha)}{\partial \alpha}\right]$	$= \frac{\partial}{\partial \alpha} F(s,\alpha)$
Time Multiplication	$L[tf(t)]$	$= \frac{-dF(s)}{ds}$

Transform Pairs

Function	Laplace Transform
1	$\frac{1}{s}$
$H(t-T) = 0 \quad t < T$ $= 1 \quad t > T$	$\frac{1}{s} e^{-sT}$
t^n	$\frac{n!}{s^{n+1}}$
e^{-at}	$\frac{1}{s+a}$
$\sin \omega t$	$\frac{\omega}{s^2+\omega^2}$
$\cos \omega t$	$\frac{s}{s^2+\omega^2}$
$1 - e^{-t/T}$	$\frac{1}{s(1+Ts)}$
$\frac{\omega_n}{\sqrt{1-\xi^2}} e^{-\xi\omega_n t} \sin\left(\omega_n\sqrt{1-\xi^2}t\right)$	$\frac{1}{1 + 2\frac{\xi s}{\omega_n} + \frac{s^2}{\omega_n^2}}$

$$1 - \frac{1}{\sqrt{1-\xi^2}} e^{-\xi\omega_n t} \sin\left(\omega_n\sqrt{1-\xi^2}\,t + \cos^{-1}\xi\right) \qquad \frac{1}{s\left(1 + 2\frac{\xi s}{\omega_n} + \frac{s^2}{\omega_n^2}\right)}$$

4.10 Numerical analysis

(i) Approximate solution of an algebraic equation $f(x) = 0$

(a) Newton's Method

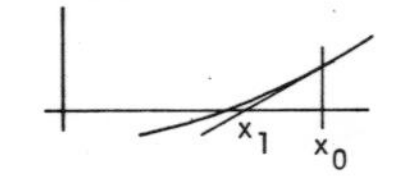

$$x_1 = x_0 - \frac{f(x_0)}{f'(x_0)}$$

(b) Secant Method

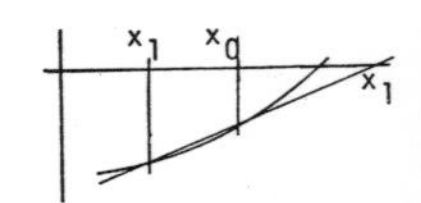

$$x_1 = \frac{-x_0\, f(x_{-1}) + x_{-1}\, f(x_0)}{f(x_0) - f(x_{-1})}$$

(ii) Least-squares fitting of a straight line

If y_i $(i = 1, 2, \ldots n)$ are the experimentally observed values of y at chosen (exact) values of x_i of the variable x, the line of 'best fit' passes through the centroid

$$\bar{x} = \frac{1}{n}\sum_{i=1}^{n} x_i \qquad \bar{y} = \frac{1}{n}\sum_{i=1}^{n} y_i$$

and is given by $y = mx + c$ where,

$$m = \frac{\Sigma(x_i-\bar{x})(y_i-\bar{y})}{\Sigma(x_i-\bar{x})^2}, \quad c = \bar{y} - m\bar{x}$$

$$= \frac{\Sigma x_i y_i - n\bar{x}\bar{y}}{\Sigma x_i^2 - n\bar{x}^2}$$

(iii) Finite-difference formulae

$$\Delta f(x) = f(x + h) - f(x)$$

$$f'(x) = \frac{f(x+h) - f(x-h)}{2h} + O(h^2)$$

$$f''(x) = \frac{f(x+h) - 2f(x) + f(x-h)}{h^2} + O(h^2)$$

$$f'''(x) = \frac{f(x+2h) - 2f(x+h) + 2f(x-h) - f(x-2h)}{2h^3}$$

(iv) Lagrange's interpolation formula for unequal intervals.

The polynomial $P(x)$ of degree 2 passing through the three points (x_i, y_i), $i = 1, 2, 3$, is

$$P(x) = \frac{(x-x_2)(x-x_3)}{(x_1-x_2)(x_1-x_3)} y_1 + \frac{(x-x_1)(x-x_3)}{(x_2-x_1)(x_2-x_3)} y_2 + \frac{(x-x_1)(x-x_2)}{(x_3-x_1)(x_3-x_2)} y_3$$

(v) Formulae for numerical integration

Equal intervals h

$$x_n = x_0 + nh , \qquad y_n = y(x_n)$$

(a) Trapezoidal Rule (1-strip):

$$\int_{x_0}^{x_1} y(x)dx = \frac{h}{2}[y_0 + y_1] + \varepsilon ,$$

$$\varepsilon \simeq \frac{-h^3}{12} y''_0 \text{ or, } \frac{-h}{12} \Delta^2 y_0$$

(b) Simpson's Rule (2-strip):

$$\int_{x_0}^{x_2} y(x)dx = \frac{h}{3}[y_0 + 4y_1 + y_2] + \varepsilon,$$

$$\varepsilon \simeq \frac{-h^5}{90} y_1^{(4)} \text{ , or } \frac{-h}{90} \Delta^4 y_0$$

(vi Runge-Kutta

2nd order: $y_{n+1} = y_n + \frac{h}{2}\left\{f(x_n, y_n) + f(x_n+h, y_n+k_1)\right\}$

4th order: $y_{n+1} = y_n + \frac{1}{6}(k_1+2k_2+2k_3+k_4)$

$k_1 = hf(x_n, y_n)$

$k_2 = hf\left(x_n+\frac{h}{2}, y_n+\frac{k_1}{2}\right)$

$k_3 = hf\left(x_n+\frac{h}{2}, y_n+\frac{k_2}{2}\right)$

$k_4 = hf(x_n+h, y_n+k_3)$

5. ANALYSIS OF EXPERIMENTAL DATA

5.1 Probability distributions for discrete random variables

Notation: $P(r) = f(r) \Longrightarrow$ the probability distribution of random variable r is $f(r)$

μ = mean value of r = $\sum_{i=1}^{N} r_i f(r_i)$

σ^2 = variance of r = $\sum_{i=1}^{N} r_i^2 f(r_i) - \mu^2$

$\binom{n}{r}$ = binomial coefficient = $\frac{n!}{(n-r)!r!} = \binom{n}{n-r}$

evaluate using Pascal's Triangle

	r = 0	1	2	3	4	5	6	7	8	9	10
n = 0	1										
1	1	1									
2	1	2	1								
3	1	3	3	1							
4	1	4	6	4	1						
5	1	5	10	10	5	1					
6	1	6	15	20	15	6	1				
7	1	7	21	35	35	21	7	1			
8	1	8	28	56	70	56	28	8	1		
9	1	9	36	84	126	126	84	36	9	1	
10	1	10	45	120	210	252	210	120	45	10	1

(a) Binomial:

n = number of trials with constant probability p of success in each

r = number of successes

$P(r) = \binom{n}{r} p^r (1-p)^{n-r}$ $\quad r = 0, 1, 2, \ldots n$

$\mu = np$, $\quad \sigma^2 = np(1-p)$

(b) Poisson:

μ = mean rate of occurrence of an event

r = number of events actually occurring in unit time

$P(r) = e^{-\mu} \mu^r / r!$ $\quad r = 0, 1, \ldots$

$\sigma^2 = \mu$

5.2 <u>Probability distributions for continuous random variables</u>

(a) <u>Exponential</u>:

probability density function $f(x) = \lambda e^{-\lambda x}$,

$x \geqslant 0, \lambda > 0$

$\mu = 1/\lambda$ $\sigma^2 = 1/\lambda^2$

(b) <u>Normal</u>: the standardised normal distribution, $N(0,1)$ has probability density function

$$\phi(z) = \frac{1}{\sqrt{(2\pi)}} e^{-\frac{1}{2}z^2}$$

$\mu = 0$ $\sigma = 1$

Φ = cumulative distribution function

$\Phi(z)$ = probability that the random variable is observed to have a value $\leqslant z$ (the shaded area shown)

$$\Phi(z) = \int_{-\infty}^{z} \frac{1}{\sqrt{2\pi}} e^{-\frac{1}{2}t^2} dt$$

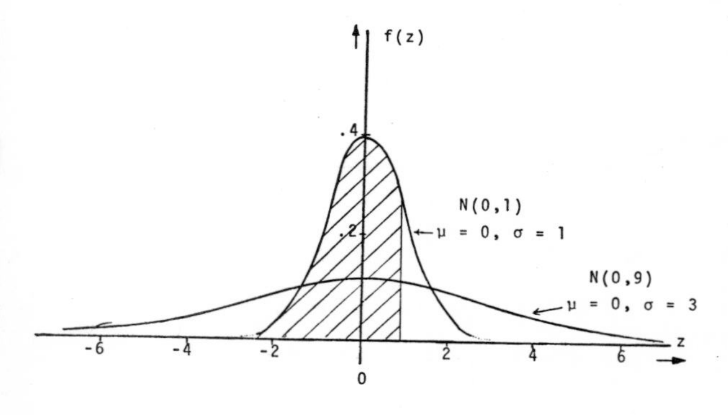

For negative z use $\Phi(-z) = 1 - \Phi(z)$

z	$\Phi(z)$	z	$\Phi(z)$	z	$\Phi(z)$
0.0	.5000	1.0	.8413	2.0	.9772
.1	.5398	.1	.8643	.1	.9821
.2	.5793	.2	.8849	.2	.9861
.3	.6179	.3	.9032	.3	.9893
.4	.6554	.4	.9192	.4	.9918
0.5	.6915	1.5	.9332	2.5	.9938
.6	.7257	.6	.9452	.6	.9953
.7	.7580	.7	.9554	.7	.9965
.8	.7881	.8	.9641	.8	.9974
.9	.8159	.9	.9713	.9	.9981
				3.0	.9987
				4.0	.99997

Percentage points of the Normal Distribution N(0,1)

$\Phi(z)$	%(1-tail)	%(2-tails)	z
.9500	5.0	10	1.6449
.9750	2.5	5	1.9600
.9900	1.0	2	2.3263
.9950	0.5	1	2.5758

The general normal distribution $N(\mu,\sigma^2)$ has probability density function $f(x) = \frac{1}{\sigma\sqrt{(2\pi)}} e^{-(x-\mu)^2/2\sigma^2}$, $-\infty < x < \infty$

where $\int_{-\infty}^{\infty} f(x)dx = 1$

and cumulative distribution function F(x)

$$F(x) = \int_{-\infty}^{(x-\mu)/\sigma} \frac{1}{\sqrt{2\pi}} e^{-\frac{1}{2}u^2} du = \Phi\left(\frac{x-\mu}{\sigma}\right)$$

To use tables of $\Phi(z)$, take $z = \frac{x-\mu}{\sigma}$.

5.3 Experimental Samples

$x_1, x_2, \ldots x_n$ denote a set of n observations of a random variable having a normal distribution whose population mean μ is unknown.

Range $= x_{max} - x_{min}$

Sample mean $m = \frac{1}{n} \Sigma x_i$

Average deviation $= \frac{1}{n} \Sigma |x_i - m|$

Sample standard deviation $= s$,

Sample variance $= s^2 = \frac{1}{n-1} \Sigma (x_i - m)^2$

Distribution of x is $N(\mu, \sigma^2)$

Distribution of m is $N(\mu, \sigma^2/n)$

Distribution of $\frac{m-\mu}{(\sigma/\sqrt{n})}$ is $N(0,1)$

i.e. standard error of sample means $= \frac{\sigma}{\sqrt{n}}$

If population variance σ^2 is known,

95% confidence interval for μ is $m \pm 1.96\ \sigma/\sqrt{n}$

99% " " " " " $m \pm 2.58\ \sigma/\sqrt{n}$

If population variance σ^2 is unknown: $\frac{m-\mu}{s/\sqrt{n}}$ has the t-distribution with $n-1$ degrees of freedom (t_{n-1}) and the 95% confidence interval for μ is obtained from $m \pm t_c\ s/\sqrt{n}$ and the table.

95% points of the t-distribution

n-1	t_c	n-1	t_c	n-1	t_c
1	12.7	6	2.45	12	2.18
2	4.30	7	2.36	15	2.13
3	3.18	8	2.31	20	2.09
4	2.78	9	2.26	30	2.04
5	2.57	10	2.23	60	2.00
				∞	1.96

Thus for $n > 20$, $m \pm 1.96\ s/\sqrt{n}$ is a good approximation to the population mean with a 95% confidence.

5.4 Combination of Errors

If results are Normally Distributed, the Most Probable Error S_z in the calculated result $z = f(x, y, etc)$, due to the independent standard errors S_x, S_y, etc. in x, y, etc. is given by,

$$(S_z)^2 = \left(\frac{\partial z}{\partial x} S_x\right)^2 + \left(\frac{\partial z}{\partial y} S_y\right)^2 + \ldots \text{ etc.}$$

If the function f consists of multiplied and divided terms ONLY (i.e. no addition or subtraction)

$$\left(\frac{S_z}{z}\right)^2 = \left(n \frac{S_x}{x}\right)^2 + \left(m \frac{S_y}{y}\right)^2 + \ldots \text{ etc.}$$

where n, m, etc. are the powers of x, y, etc. in f.

Notes

(1) The Maximum Possible Error ($\delta z = \frac{\partial z}{\partial x} \delta x + \frac{\partial z}{\partial y} \delta y$, etc.) is rarely of interest in engineering

(2) Instrument 'rounding off' error $\pm\delta x$ may be treated as a Normally Distributed error by the equivalence $S_x \triangleq \frac{2}{3} \delta x$.

6. MECHANICS

Moments of inertia and Second moments of area - General theorems

N.B. The symbol I is used for both second moment of area and moment of inertia.

(i) Parallel axis theorem: Solids or laminae

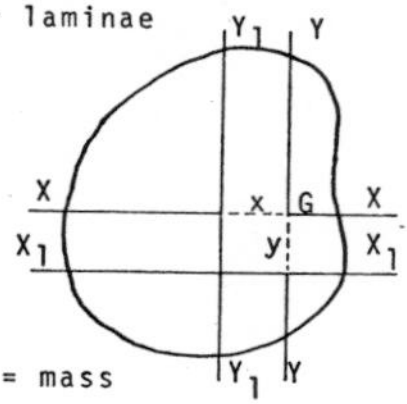

Centroid is at G

Centre of mass is at G

$$I_{X_1X_1} = I_{XX} + Cy^2$$

$$I_{Y_1Y_1} = I_{YY} + Cx^2$$

where for moment of inertia C = mass

and for second moment of area of a lamina C = area

(ii) Perpendicular axis theorem for laminae:

Polar second moment $J_o = I_{XX} + I_{YY} = I_{ZZ}$

Radii of gyration k

Second moment of area $I = Ak^2$

Moment of inertia $I = mk^2$

where A = area, m = mass.

(a) UNIFORM ROD	k^2_{XX}	k^2_{YY}	A
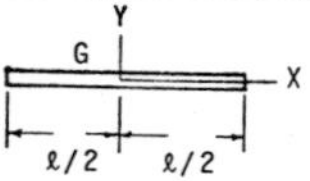	-	$\frac{\ell^2}{12}$	-
(b) LAMINAE			
Rectangle 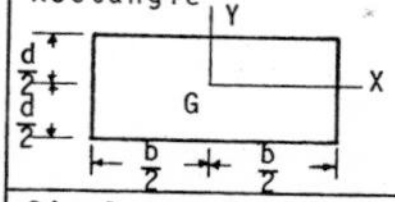	$\frac{1}{12}d^2$	$\frac{1}{12}b^2$	db
Circle 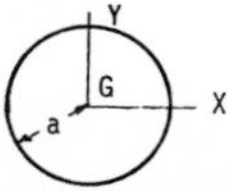	$\frac{1}{4}a^2$	$\frac{1}{4}a^2$	πa^2

(b) LAMINAE (cont)	k_{XX}^2	k_{YY}^2	A
Semi-circle 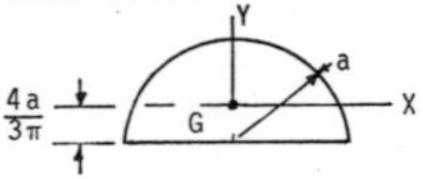	$a^2 \left\| \frac{1}{4} - (\frac{4}{3\pi})^2 \right\|$	$\frac{1}{4} a^2$	$\frac{\pi a^2}{2}$
Triangle 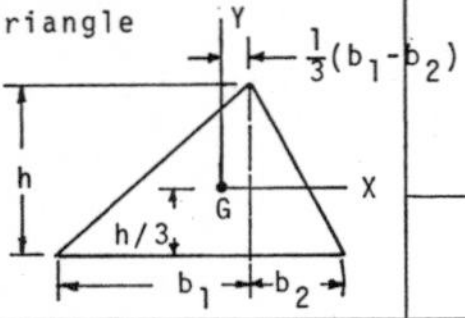	$\frac{1}{18} h^2$	$\frac{1}{18} (b_1^2 + b_1 b_2 + b_2^2)$	$\frac{h}{2}(b_1 + b_2)$
	$k_{XY}^2 = \frac{1}{36} h(b_1 - b_2)$		
Ellipse 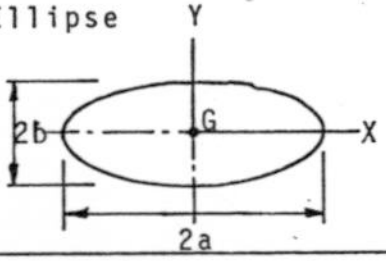	$\frac{b^2}{4}$	$\frac{a^2}{4}$	πab

(c) SOLIDS	k_{XX}^2	k_{YY}^2 and k_{ZZ}^2	V
Cylinder 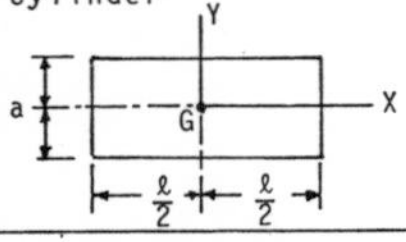	$\frac{1}{2} a^2$	$\frac{1}{4} a^2 + \frac{1}{12} \ell^2$	$\pi a^2 \ell$
Thin-walled cylinder 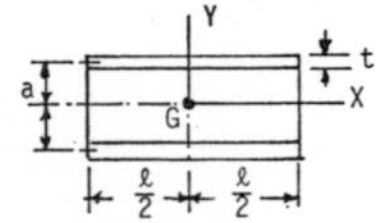	$a^2 + \frac{1}{4} t^2$	$\frac{1}{2} a^2 + \frac{1}{8} t^2 + \frac{1}{12} \ell^2$	$2\pi a t \ell$
Thick walled cylinder 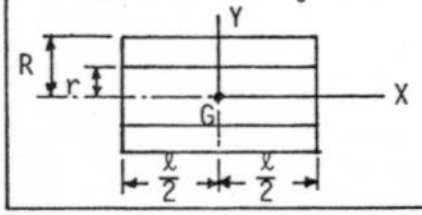	$\frac{1}{2}(R^2 + r^2)$	$\frac{\ell^2}{12} + \frac{R^2 + r^2}{4}$	$\pi(R^2 - r^2)\ell$

(c) SOLIDS (cont)	k_{XX}^2 and k_{ZZ}^2	k_{YY}^2	V
Sphere	$\frac{2}{5}a^2$	$\frac{2}{5}a^2$	$\frac{4}{3}\pi a^3$
Cone	$\frac{3(4a^2+h^2)}{80}$	$\frac{3a^2}{10}$	$\frac{\pi}{3}a^2h$

Mohr's Circle for Second Moment of Area

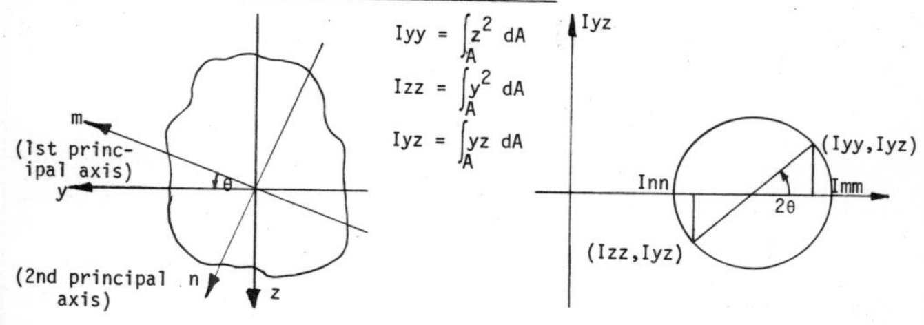

Constant acceleration equations

$$v = u + at$$
$$v^2 = u^2 + 2ax$$
$$x = ut + \frac{1}{2}at^2$$

Friction

coefficient of static friction $\mu = \tan\phi$

for no slipping $\frac{F}{N} \leqslant \mu$

Accelerations due to rotation

Coriolis $= 2\,\underline{\omega} \times \left(\frac{\partial \underline{r}}{\partial t}\right)$

Central $= \underline{\omega} \times (\underline{\omega} \times \underline{r})$

DRY SLIDING FRICTION COEFFICIENTS

Clutches	0.3-0.4
Brakes (lining)	0.35-0.5
" (pads)	~0.3
Nylon/Steel	0.3-0.5
Filled PTFE/Steel	0.05-0.3
Perspex/Steel	~0.5
Rubber/Steel	0.6-0.9
Rubber/Asphalt	0.5-0.8
Lignum vitae/Steel	~0.1

7. PROPERTIES AND MECHANICS OF SOLIDS

7.1 Bonding

(a) Condon-Morse Equation $V_{total} = \frac{-Ae^2}{r^n} + \frac{B}{r^m} + C$

(b) Ionic Bond Equation $V_o = \frac{-Z_1^* Z_2^* e^2}{4\pi d\varepsilon_o}(1 - \frac{1}{n}) + \Delta E$

(c) Theoretical Density $\rho = \frac{nA}{VN}$

7.2 Atomic sizes in substitutional alloys

Element	Seitz radius r_o(Å) (at 20°C)	Effective valency in solution
Al	1.58	3
Au	1.59	1
Cu	1.41	1
Fe(α)	1.41	?
Mg	1.85	2
Ni	1.38	1
P	1.58	3
Pb	1.95	4
Si	1.67	4
Sn	1.86	4
Zn	1.54	2

7.3 Phase Transformations

Length and volume changes may be related by:-

$$(1 + \Delta V/V) = (1 + \Delta L/L)^3$$

7.4 Crystallography

(a) In the Miller system:

Specific Plane	(h.k.ℓ)
Family of Planes	{h.k.ℓ}
Specific Direction	[h.k.ℓ]
Family of Directions	<h.k.ℓ>

(b) Inter-planar spacings for Cubics

$$d_{(h.k.\ell)} = \frac{a}{\sqrt{h^2+k^2+\ell^2}} = \frac{a}{\sqrt{N}}$$

(c) Quadratic Forms of Miller Indicies (N values)

Cubic Structure	N values
Simple	1,2,3,4,5,6,8,9,10,11,12,13,14,16,17,18,19,20.....
Face Centred	3,4,8,11,12,16,19,20,24,27,32.....
Body Centred	2,4,6,8,10,12,14,16,18,20,22,24,26,30.....
Diamond	3,8,11,16,19,24,27,32....

7.5 Defects and Diffusion Data

(a) Number of Defects $n = \alpha N e^{\frac{-Q}{kT}}$

(b) Diffusivity $D = D_0 e^{\frac{-Q}{kT}}$

Note: these equations may be expressed in terms of R_0 rather than k, the value of Q must be quoted in the appropriate units.

(c) Macroscopic Diffusion

(i) D constant with composition, $\frac{dc}{dx}$ constant with time

$$J = - D \frac{dc}{dx}$$ (This is a special case of (ii))

(ii) D constant with composition, $\frac{dc}{dx}$ varies with time

$$\frac{dc}{dt} = D \frac{d^2c}{dx^2}$$

Solution for a constant surface potential and impermeable sides

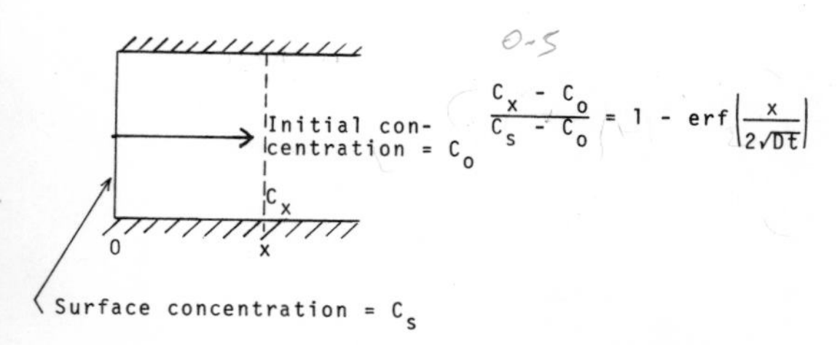

$$\frac{C_x - C_0}{C_s - C_0} = 1 - \operatorname{erf}\left(\frac{x}{2\sqrt{Dt}}\right)$$

7.6 Selected Values of Error Function $\frac{2}{\sqrt{\pi}}\int_0^z e^{-u^2}du$

z	erfz	z	erfz	z	erfz	z	erfz
0.00	0.0000	0.68	0.6638	1.36	0.9456	2.00	0.9953
0.02	0.0226	0.70	0.6778	1.38	0.9490	2.05	0.9963
0.04	0.0451	0.72	0.6914	1.40	0.9523	2.10	0.9970
0.06	0.0676	0.74	0.7047	1.42	0.9554	2.15	0.9976
0.08	0.0901	0.76	0.7175	1.44	0.9583	2.20	0.9981
0.10	0.1125	0.78	0.7300	1.46	0.9611	2.25	0.9983
0.12	0.1348	0.80	0.7421	1.48	0.9637	2.30	0.9989
0.14	0.1570	0.82	0.7538	1.50	0.9661	2.35	0.9991
0.16	0.1790	0.84	0.7651	1.52	0.9684	2.40	0.9993
0.18	0.2009	0.86	0.7761	1.54	0.9706	2.45	0.9995
0.20	0.2227	0.88	0.7867	1.56	0.9726	2.50	0.9996
0.22	0.2443	0.90	0.7969	1.58	0.9746	2.55	0.9997
0.24	0.2657	0.92	0.8068	1.60	0.9764	2.60	0.9998
0.26	0.2869	0.94	0.8163	1.62	0.9780	2.65	0.9998
0.28	0.3079	0.96	0.8254	1.64	0.9796	2.70	0.9999
0.30	0.3286	0.98	0.8342	1.66	0.9811	2.75	0.9999
0.32	0.3491	1.00	0.8427	1.68	0.9825	2.80	0.9999
0.34	0.3694	1.02	0.8508	1.70	0.9838	2.85	0.9999
0.36	0.3893	1.04	0.8587	1.72	0.9850	2.90	1.0000
0.38	0.4090	1.06	0.8661	1.74	0.9861	2.95	1.0000
0.40	0.4284	1.08	0.8733	1.76	0.9872	3.00	1.0000
0.42	0.4475	1.10	0.8802	1.78	0.9882	4.00	1.0000
0.44	0.4662	1.12	0.8868	1.80	0.9891		
0.46	0.4847	1.14	0.8931	1.82	0.9899		
0.48	0.5028	1.16	0.8991	1.84	0.9907		
0.50	0.5205	1.18	0.9048	1.86	0.9915		
0.52	0.5379	1.20	0.9103	1.88	0.9922		
0.54	0.5549	1.22	0.9155	1.90	0.9928		
0.56	0.5716	1.24	0.9205	1.92	0.9934		
0.58	0.5879	1.26	0.9252	1.94	0.9939		
0.60	0.6039	1.28	0.9297	1.96	0.9944		
0.62	0.6194	1.30	0.9340	1.98	0.9949		
0.64	0.6346	1.32	0.9381				
0.66	0.6494	1.34	0.9419				

7.7 Fracture

i) Fatigue

Manson-Coffin Law $\sqrt{Np}\ \varepsilon p = c$ (c = constant)

Miner's Rule $\sum\left(\frac{ni}{Ni}\right) = 1$

Rayleigh Distribution $P(\sigma) = \sigma r^{-2} \exp\left\{-\frac{1}{2}\left(\frac{\sigma}{r}\right)^2\right\}$

Fraction of peak exceeding stress (σ) expressed in terms of E $E(\sigma) = \exp\left\{-\frac{1}{2}\left(\frac{\sigma}{r}\right)^2\right\}$

ii) Fracture Toughness

Stress Intensity $K = Q\sigma\sqrt{\pi a}$

Paris Equation $\frac{da}{dN} = A(\Delta K)^n = A_1 a^{m/2}$ (A, A_1, n and m are constants)

7.8 Some typical values of physical properties

All values are given, unless otherwise stated, for a temperature of $20^{\circ}C$.

	Carbon Steel	Aluminium Alloys	Brass 65/35	Copper	Concrete	Stainless Steels	Wood
ρ (kg/m^3)	7850	2720	8450	8960	2400	8000	400-800
E (GN/m^2)	207	68.9	105	104	13.8	213	8-13
G (GN/m^2)	79.6	26.5	38.0	46		82	
K (GN/m^2)	172	57.5	115	130		178	
ν	0.3	0.3	0.35	0.35	0.1	0.3	
α (μm/(mK))	11	23	19	11.2		18	~0.15
σ_y (MN/m^2)	230-460	30-280	62-430	47-320		200-585	
σ_f (MN/m^2)	400-770	90-300	330-530	200-350	27-55	500-800	50-100

K for water is 2.3 GN/m^2

The lower values of σ_y and σ_f for carbon and stainless steels refer to materials such as plates and tubes while the higher figures refer to heat-treated material such as used for bolts. The range of values for aluminium, copper and brass is due to the change in material property achieved by heat-treatment and/or mechanical work.

Metals

Property	Copper	Iron
Crystal structure	f.c.c.	b.c.c.
Bonding	metallic	metallic
Lattice constant (Å)	3.61	2.86
Atomic volume (m^3/kg mol)	7.09×10^{-3}	7.10×10^{-3}
ρ (kg/m^3)	8.96×10^3	7.87×10^3
Resistivity (Ω m)	1.72×10^{-8}	10×10^{-8}
Cohesive energy (J/kg mol)	3.38×10^8	4.05×10^8
Melting point (^oC)	1083	1530
α(μm/(mK))	16.7	12.1
Fermi energy (eV)	7.04	11.2
Work function (eV)	4.07 - 4.18	3.91 - 4.77
Temperature coefficient of resistance (K^{-1})	+0.0043	+0.0065
Effective radius (Å) of		
(a) neutral atom	1.27	1.26
(b) singly charged ion	0.96	-
(c) doubly charged ion	0.70	0.75

Semiconductors

Property	Germanium	Silicon
Crystal structure	diamond	diamond
Bonding	covalent	covalent
Lattice constant (Å)	5.6575	5.4307
Atomic volume (m^3/kg mol)	13.5×10^{-3}	12.0×10^{-3}
Density (kg/m^3)	5.32×10^3	2.33×10^3
Cohesive energy (J/kg mol)	3.72×10^8	4.39×10^8
Melting point (^oC)	958.5	1412
Mobility (m^2/(V s))	electrons 0.38 holes 0.18	electrons 0.19 holes 0.05
Energy gap (eV) (room temperature)	0.67	1.107
Density of states effective mass	electrons 0.35 m_e holes 0.56 m_e	electrons 0.58 m_e holes 1.06 m_e
α(μm/(mK))	5.75	7.6

Polymers

PROPERTY	Polyethylene (H.D)	Polyvinyl Chloride	Polystyrene
Polymer Structure	$[\bullet{-}CH_2{-}CH_2{-}\bullet]_n$	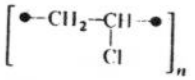 $[\bullet{-}CH_2{-}CH(Cl){-}\bullet]_n$	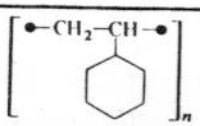 $[\bullet{-}CH_2{-}CH(C_6H_5){-}\bullet]_n$
Structural State	Crystalline	Amorphous/ Slightly Crystalline	Amorphous/ Crystalline
$\rho(kg/m^3)$	0.96×10^3	1.7×10^3	1.05×10^3
Resistivity (Ωm)	$10^6 - 10^{10}$	10^5	10^{10}
$\alpha(\mu m/mK)$	120	190	63
$E(GN/m^2)$	70-280	2500-3500	3500-4200
$\sigma_f(MN/m^2)$	7-14	28-40	35-50
Tg(K)	153	353	373

PROPERTY	Polymethylmethacrylate	Polytetrafluoroethylene	Polyisoprene (Natural Rubber)
Polymer Structure	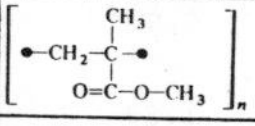 $[\bullet{-}CH_2{-}C(CH_3)(C(=O){-}O{-}CH_3){-}\bullet]_n$	$[\bullet{-}CF_2{-}CF_2{-}\bullet]_n$	$[\bullet{-}CH_2{-}C(CH_3){=}CH{-}CH_2{-}\bullet]_n$
Structural State	Amorphous	Crystalline	Elastomer
$\rho(kg/m^3)$	1.2×10^3	2.2×10^3	1.5×10^3
Resistivity (Ωm)	10^8	10^8	$10^5 - 10^7$
$\alpha(\mu m/mK)$	90	100	-
$E(GN/m^2)$	2500-4000	400-650	7-70
$\sigma_f(MN/m^2)$	50-70	14-30	2-10
Tg(K)	380	399	203

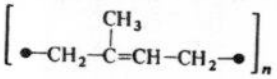

PROPERTY	Nylon 6:6	Phenol-Formaldehyde Resin (Bakelite)
Polymer Structure	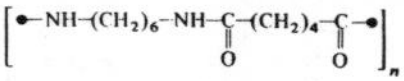 $[\bullet{-}NH{-}(CH_2)_6{-}NH{-}C(=O){-}(CH_2)_4{-}C(=O){-}\bullet]_n$	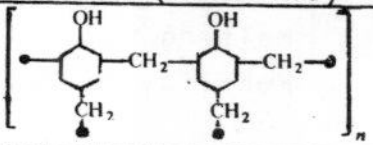 $[\bullet{-}C_6H_2(OH)(CH_2\bullet){-}CH_2{-}C_6H_2(OH)(CH_2\bullet){-}CH_2{-}\bullet]_n$
Structure State	Crystalline	Amorphous
$\rho(kg/m^3)$	1.15×10^3	1.3×10^3
Resistivity (Ωm)	10^6	10^4
$\alpha(\mu m/mK)$	100	72
$E(GN/m^2)$	2000-3000	7000
$\sigma_f(MN/m^2)$	50-70	50
Tg(K)	323	-

	IA	IIA		IIIA	IVA	VA	VIA	VIIA	VIII			IB	IIB		IIIB	IVB	VB	VIB	VIIB	
1s	1 H	2 He																		
	s 1	s 2																		
2s	3 Li	4 Be												2p	5 B	6 C	7 N	8 O	9 F	10 Ne
3s	11 Na	12 Mg												3p	13 Al	14 Si	15 P	16 S	17 Cl	18 Ar
4s	19 K	20 Ca	3d	21 Sc	22 Ti	23 V	24 Cr	25 Mn	26 Fe	27 Co	28 Ni	29 Cu	30 Zn	4p	31 Ga	32 Ge	33 As	34 Se	35 Br	36 Kr
5s	37 Rb	38 Sr	4d	39 Y	40 Zr	41 Nb	42 Mo	43 Tc	44 Ru	45 Rh	46 Pd	47 Ag	48 Cd	5p	49 In	50 Sn	51 Sb	52 Te	53 I	54 Xe
6s	55 Cs	56 Ba	5d	57 La •	72 Hf	73 Ta	74 W	75 Re	76 Os	77 Ir	78 Pt	79 Au	80 Hg	6p	81 Tl	82 Pb	83 Bi	84 Po	85 At	86 Rn
					d 2 s 2	d 3 s 2 4 1 3 2	d 5 s 1 5 1 4 2	d 5 s 2 6 1 5 2	d 6 s 2 7 1 6 2	d 7 s 2 8 1 7 2	d 8 s 2 10 0 9 1	d 10 s 1	d 10 s 2		s 2 p 1	s 2 p 2	s 2 p 3	s 2 p 4	s 2 p 5	s 2 p 6
7s	87 Fr	88 Ra	6d	89 Ac •																
	s 1	s 2		d 1 s 2																

4f	58 Ce	59 Pr	60 Nd	61 Pm	62 Sm	63 Eu	64 Gd	65 Tb	66 Dy	67 Ho	68 Er	69 Tm	70 Yb	71 Lu	IIIA
4f 5s 5p 5d 6s	2 2 6 0 2	3	4	5	6	7	7 2 6 1 2	9 2 6 0 2	10	11	12	13	14	14 2 6 1 2	
		Outer sub-shells as for Ce							Outer sub-shells as for Ce & Tb						
5f	90 Th	91 Pa	92 U	93 Np	94 Pu	95 Am	96 Cm	97 Bk	98 Cf	99 Es	100 Fm	101 Md	102 No	103 Lw	
5f 6s 6p 6d 7s	0 2 6 2 2	2 2 6 1 2	3	4	5	6	7	8	9	10	11	12	13	14	
			Outer sub-shells as for Pa												

	1	2	3	4	5	6	7	
⊕	s	s p	s p d	s p d f	s p d f	s p d	s	← Principal Shell ← Sub-shell
	2	2 6	2 6 10	2 6 10 14	2 6 10 14	2 6 10	2	← Total Number of electrons required to fill each sub-shell

1. The atomic number indicates the number of protons in the nucleus of an atom. In the neutral atom these protons are electrically balanced by an equal number of electrons outside the nucleus. Only neutral atoms are considered in the Periodic Classification.
2. Electrons travel far from the nucleus but if those regions where they spend most of their time are considered, a well-defined pattern of layers or 'Principal Shells' appears. Each shell is known by a Principal Quantum Number 1, 2, 3 . . . 7 or sometimes by the letters K, L, M . . . etc.
3. In each shell the electrons move around the nucleus in complicated, three-dimensional patterns called Orbitals. The laws of Quantum Mechanics permit only certain types of orbital. An electron following one of these paths possesses an amount of energy (Energy Level) characteristic of that orbital.
4. Four types of orbital are encountered; they are identified by the letters *s p d* and *f*. *s* is the simplest whilst *p*, *d* and *f* are progressively more complex.
5. The number of orbitals per shell increases with shell number. (See the lower diagram overleaf.) The first contains only an *s* orbital, the second an *s* and three *p*'s, the third adds five *d* orbitals and the fourth seven *f*'s. These groups of like orbitals in any Principal Shell are called *s p d* or *f* sub-shells.

   Each sub-shell, depending on its principal quantum number and type, has a characteristic energy the order of which is generally proportional to the distance of the sub-shell from the nucleus.
6. Each orbital accepts either one or two electrons and the maximum number of electrons per sub-shell is shown on the diagram.
7. Electrons take positions in orbitals where the energy level is lowest. Up to element 18 (Argon) sub-shells and shells are built in an orderly sequence to maximum capacity. But in the next group the order changes because it happens that the energy level of the 4 *s* state is a little lower than that of the 3 *d* state.
8. The first transition series begins with Scandium (element 21) where the energy levels of the 4 *s* and 3 *d* orbitals are so nearly equal that there is a tendency for electrons to move from one orbital to another, causing variable valency. The same happens in the fifth period with 5 *s* and 4 *d* orbitals and in the sixth period with the 6 *s* and 5 *d* orbitals.
9. In the Lanthanide and Actinide series of elements, the 4 *f* and 5 *f* orbitals are occupied only after the *s p d* and *s* orbitals outside them have filled or begun to fill. The effect upon the chemistry of the elements is very small because the *f* orbitals are deep in the core of the atom. For this reason there is little difference between one element and its immediate neighbours.
10. In any element, the so called Valency Electrons are those moving in orbitals of the highest energy levels. In this Chart of the Periodic Classification, the number and position of the valency electrons is indicated in the boxes underneath the various columns e.g. Rhodium—element 45—has nine valency electrons; 8 in the 4 *d* sub-shell and 1 in the 5 *s*.

    The particular sub-shell being filled with electrons is shown by the figures 4 *s*, 3 *d*, 4 *p* etc. in front of the rows of elements e.g. the 3 *d* in front of elements 21—30.

NOTES

1. The following atomic weights are based on the exact number 12 for the carbon isotope 12, as agreed between the International Unions of Pure and Applied Physics and of Pure and Applied Chemistry, 1961.

2. The values given normally indicate the mean atomic weight of the mixture of isotopes found in nature. Particular attention is drawn to the value for hydrogen, boron, carbon, oxygen, silicon and sulphur, where the deviation shown is due to variation in relative concentration of isotopes.

Symbol	*Name*	*Atomic Number*	*Atomic Weight*
A or Ar	Argon	18	39·948
Ac	Actinium	89	—
Ag	Silver	47	107·870
Al	Aluminium	13	26·9815
Am	Americium	95	—
As	Arsenic	33	74·9216
At	Astatine	85	—
Au	Gold	79	196·967
B	Boron	5	10·811 ±0·003
Ba	Barium	56	137·34
Be	Beryllium	4	9·0122
Bi	Bismuth	83	208·980
Bk	Berkelium	97	—
Br	Bromine	35	79·909
C	Carbon	6	12·01115 ±0·00005
Ca	Calcium	20	40·08
Cd	Cadmium	48	112·40
Ce	Cerium	58	140·12
Cf	Californium	98	—
Cl	Chlorine	17	35·453
Cm	Curium	96	—
Co	Cobalt	27	58·9332
Cr	Chromium	24	51·996
Cs	Caesium	55	132·905
Cu	Copper	29	63·54
Dy	Dysprosium	66	162·50
Er	Erbium	68	167·26
Es	Einsteinium	99	—
Eu	Europium	63	151·96
F	Fluorine	9	18·9984
Fe	Iron	26	55·847
Fm	Fermium	100	—
Fr	Francium	87	—
Ga	Gallium	31	69·72
Gd	Gadolinium	64	157·25
Ge	Germanium	32	72·59
H	Hydrogen	1	1·00797 ±0·00001
He	Helium	2	4·0026
Hf	Hafnium	72	178·49
Hg	Mercury	80	200·59
Ho	Holmium	67	164·930
	Iodine	53	126·9044
In	Indium	49	114·82
Ir	Iridium	77	192·2
K	Potassium	19	39·102
Kr	Krypton	36	83·80
La	Lanthanum	57	138·91
Li	Lithium	3	6·939
Lu	Lutetium	71	174·97
Md	Mendeleevium	101	—
Mg	Magnesium	12	24·312
Mn	Manganese	25	54·9380
Mo	Molybdenum	42	95·94
N	Nitrogen	7	14·0067
Na	Sodium	11	22·9898
Nb	Niobium	41	92·906
Nd	Neodymium	60	144·24
Ne	Neon	10	20·183
Ni	Nickel	28	58·71
No	Nobelium	102	—
Np	Neptunium	93	—
O	Oxygen	8	15·9994 ±0·0001
Os	Osmium	76	190·2
P	Phosphorus	15	30·9738
Pa	Protoactinium	91	—
Pb	Lead	82	207·19
Pd	Palladium	46	106·4
Pm	Promethium	61	—
Po	Polonium	84	—
Pr	Praseodymium	59	140·907
Pt	Platinum	78	195·09
Pu	Plutonium	94	—
Ra	Radium	88	—
Rb	Rubidium	37	85·47
Re	Rhenium	75	186·2
Rh	Rhodium	45	102·905
Rn	Radon	86	—
Ru	Ruthenium	44	101·07
S	Sulphur	16	32·064 ±0·003
Sb	Antimony	51	121·75
Sc	Scandium	21	44·956
Se	Selenium	34	78·96
Si	Silicon	14	28·086 ±0·001
Sm	Samarium	62	150·35
Sn	Tin	50	118·69
Sr	Strontium	38	87·62
Ta	Tantalum	73	180·948
Tb	Terbium	65	158·924
Tc	Technetium	43	—
Te	Tellurium	52	127·60
Th	Thorium	90	232·038
Ti	Titanium	22	47·90
Tl	Thallium	81	204·37
Tm	Thulium	69	168·934
U	Uranium	92	238·03
V	Vanadium	23	50·942
W	Tungsten	74	183·85
Xe	Xenon	54	131·30
Y	Yttrium	39	88·905
Yb	Ytterbium	70	173·04
Zn	Zinc	30	65·37
Zr	Zirconium	40	91·22

8. THERMODYNAMICS AND FLUID MECHANICS

8.1 Thermodynamic Relationships

1st Law	$dQ - dW = dU$
Enthalpy	$H = U + pV$ or $h = u + pv$
For reversible process	$dS = \left(\frac{dQ}{T}\right)_{rev}$ or $dQ = TdS$
" " "	$dW = pdV$
Helmholtz function	$F = U - TS$ or $f = u - Ts$
Gibbs function / Gibbs free energy	$G = H - TS$ or $g = h - Ts$
From 1st Law for a homogeneous fluid	$Tds = du + pdv = dh - vdp$
Specific heat at constant volume	$c_v = \left(\frac{\partial u}{\partial T}\right)_v$
Specific heat at constant pressure	$c_p = \left(\frac{\partial h}{\partial T}\right)_p$
Specific heat ratio	$\gamma = c_p/c_v$
Reversible engine (Carnot) efficiency	$= 1 - (T_{sink}/T_{source})$
Engine indicated Power	$P_i = p_m V_s N_c$
Steady flow energy equation	$(Q-W)/m = h_2 - h_1 + \frac{1}{2}(c_2^2 - c_1^2) + g(z_2 - z_1)$
Continuity equation	$\dot{m} = \rho Ac$

General relationships for a perfect gas:

$pv = RT$	$pV = mRT$
$pv_o = R_oT$	$MR = R_o$
$\Delta U = mc_v(T_2 - T_1)$	$\Delta H = mc_p(T_2 - T_1)$
$\Delta S = mc_v \ln\left(\frac{p_2}{p_1}\right) + mc_p \ln\left(\frac{v_2}{v_1}\right)$	
$c_p - c_v = R$	$\frac{\gamma - 1}{\gamma} = \frac{R}{c_p}$

Van der Waals' equation $(p + \frac{a}{v^2})(v - b) = RT$

$S = k \ln P$ $\qquad k = R_o/N$

Availability, (closed system): $(A_1 - A_o) = (U_1 + p_oV_1 - T_oS_1) - (U_o + p_oV_o - T_oS_o)$

" (flow process) : $(B_1 - B_o) = (H_1 - T_oS_1) - (H_o - T_oS_o)$

Maximum work of a Reaction $W_{max} = G_{react} - G_{prod} = R_oT \ln(K_p . p^n)$

For reversible polytropic (pV^n = constant) closed system:

$$W = (p_1V_1 - p_2V_2)/(n-1)$$

For perfect gas also:

$$W = mR(T_1 - T_2)/(n-1) \qquad \frac{T_2}{T_1} = \left(\frac{p_2}{p_1}\right)^{\frac{n-1}{n}} = \left(\frac{V_1}{V_2}\right)^{n-1}$$

$$Q = \frac{\gamma - n}{\gamma - 1}.W$$

For adiabatic reversible (isentropic reversible):

$$n = \gamma$$

For isothermal reversible:

$$W = Q = pV\ \ln\left(\frac{V_1}{V_2}\right) \quad (n = 1)$$

Maxwell relations

$$\left(\frac{\partial T}{\partial v}\right)_s = -\left(\frac{\partial p}{\partial s}\right)_v \qquad \left(\frac{\partial T}{\partial p}\right)_s = \left(\frac{\partial v}{\partial s}\right)_p$$

$$\left(\frac{\partial p}{\partial T}\right)_v = \left(\frac{\partial s}{\partial v}\right)_T \qquad \left(\frac{\partial v}{\partial T}\right)_p = -\left(\frac{\partial s}{\partial p}\right)_T$$

Heat transfer

Conduction (one dimensional) $\dot{Q}/A = -k\,dT/dx = k\left(T_1 - T_2\right)/x_{1,2}$

" (radial flow) $\dot{Q}/\ell = 2\pi k\Delta T/\ln(r_2/r_1)$

Forced convection in a tube $Nu = .023\ Re^{0.8}Pr^{0.4}$
(characteristic length = hydraulic mean diameter)(see 8.5)

Log. mean temperature difference $\dfrac{\Delta T_{in} - \Delta T_{out}}{\ln(\Delta T_{in}/\Delta T_{out})} = \Delta T_m$

Stefan-Boltzmann Law of radiation $q_b = \sigma T^4$

Radiation exchange:

Grey body to black or large enclosure $\dot{Q}/A = \sigma\varepsilon_1\left(T_1^4 - T_2^4\right)$

Large parallel grey surfaces $\dot{Q}/A = \dfrac{\sigma\left(T_1^4 - T_2^4\right)}{1/\varepsilon_1 + 1/\varepsilon_2 - 1}$

Heat transfer coefficient $h = \dot{Q}/A\Delta T$

emissivity $\varepsilon = q/q_b$

Fluid Mechanics

Statics

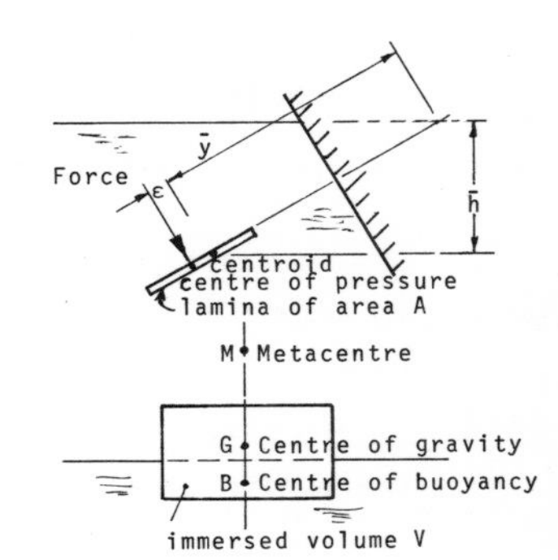

$$\frac{\partial p}{\partial z} = -g\rho$$

$$\text{Force} = g\rho A\bar{h}$$

$$\varepsilon = \frac{(Ak^2)\text{centroid}}{A\bar{y}}$$

$$\overline{GM}_{roll} = \frac{Ak^2}{V} - \overline{BG}$$

(Ak^2 is 2nd moment of area about rolling axis)

Dynamics

For simple Newtonian flow $\tau = \mu \frac{dV}{dy}$

Euler's equation $\frac{1}{\rho}\frac{dp}{dx} + c\frac{dc}{dx} + g\frac{dz}{dx} = 0$

Bernoulli's equation $\frac{p}{\rho g} + \frac{c^2}{2g} + z = \text{constant}$

For constant area flow with friction (Fanno)

$$\frac{dp}{\rho} + c\ dc + 2\frac{fc^2}{D}\ d\ell = 0$$

Acceleration along a stream-line $a_s = V_s \frac{\partial V_s}{\partial s} + \frac{\partial V_s}{\partial t}$

Acceleration normal to a streamline $a_n = \frac{V_s^2}{r} + \frac{\partial V_n}{\partial t}$

Reynolds' Equation for bearings:

$$\frac{\partial}{\partial x}\left(\frac{\rho h^3}{12\eta}\frac{\partial p}{\partial x}\right) + \frac{\partial}{\partial y}\left(\frac{\rho h^3}{12\eta}\frac{\partial p}{\partial y}\right) = \frac{1}{2}\frac{\partial}{\partial x}(\rho U h) + \frac{\partial}{\partial t}(\rho h) + \rho w$$

$$\frac{1}{r}\frac{\partial}{\partial r}\left(\frac{\rho r h^3}{12\eta}\frac{\partial p}{\partial r}\right) + \frac{1}{r^2}\frac{\partial}{\partial \phi}\left(\frac{\rho h^3}{12\eta}\frac{\partial p}{\partial \phi}\right) = \frac{1}{2}\frac{\partial}{\partial \phi}(\rho h \omega) + \frac{\partial}{\partial t}(\rho h) + \rho w$$

Hydraulic machines

Head coefficient $\psi = Y/\omega^2 D^2$

Flow " $\phi = Q/\omega D^3$

Dimensionless specific speed $n_s = \left| \omega Q^{\frac{1}{2}}/Y^{\frac{3}{4}} = \phi^{\frac{1}{2}}/\psi^{\frac{3}{4}} = \frac{\omega(\text{Power})^{\frac{1}{2}}}{\rho^{\frac{1}{2}}\ Y^{5/4}} \right|_{\eta_{MAX}}$

Dimensionless diameter $\Delta = DY^{\frac{1}{4}}/Q^{\frac{1}{2}}$

Dimensionless suction specific speed $N_{ss} = \omega Q^{\frac{1}{2}}/(NPSE)^{\frac{3}{4}}$

Cavitation number σ or $k = (P_\infty - P_v)/(\frac{1}{2}\rho V_\infty^2)$, suffix ∞, reference condition.

Cavitation number (Thoma) $\sigma_{Th} = (P_1 - P_v)/(P_2 - P_1)$

suffix 1, abs. pressure at lp side of machine; suffix 2, abs pressure at hp side of machine; suffix v, vapour pressure

Open Channel Hydraulics

Chézy equation $V = C\sqrt{RS}$

Manning equation: $V = \frac{1}{n} R^{2/3} S^{1/2}$

Steady gradually varied flow equation: $\frac{dd}{dx} = \frac{S_o - S_f}{1 - \alpha\frac{V^2}{2g}}$ (rectangular channel)

Unsteady gradually varied flow equation: $\frac{\partial d}{\partial x} + \frac{V}{g}\frac{\partial V}{\partial x} + \frac{1}{g}\frac{\partial V}{\partial t} = S_o - S_f$

Continuity equation: $A\frac{\partial V}{\partial x} + V\frac{\partial A}{\partial x} + T\frac{\partial d}{\partial t} = 0$

(no local inflow or outflow)

Conjugate depths in hydraulic jump: $\frac{d_2}{d_1} = \frac{1}{2}\left(\sqrt{1+8F_1^2} - 1\right)$ (rectangular channel)

High speed gas flow

Nozzles:

Mass flow given by $\dot{m} = AC_d\sqrt{\frac{2n}{(n-1)} p_o \rho_o\left[\left(\frac{p}{p_o}\right)^{2/n} - \left(\frac{p}{p_o}\right)^{\frac{n+1}{n}}\right]}$

Critical pressure ratio $\frac{p^*}{p_o} = \left(\frac{2}{n+1}\right)^{\frac{n}{n-1}}$

Sonic velocity $a = \sqrt{np/\rho}$

where $n \simeq 1.3$ for steam, initially superheated

$\simeq 1.135$ for steam, initially wet or dry saturated

$= \gamma \simeq 1.4$ for air

For perfect gas:

Stagnation temperature $T_o = T\left[1 + \frac{(\gamma-1)}{2} M^2\right]$

For air in isentropic flow ($\gamma = 1.4$) $\frac{\dot{m}\sqrt{T_o}}{A^* p_o} = 0.0404 \frac{kgK^{\frac{1}{2}}}{Ns}$

Turbine Ellipse Law $\frac{\dot{m}\sqrt{T_o}}{Ap_o} = \left[1 - \left(\frac{p}{p_o}\right)^2\right]^{\frac{1}{2}}$

8.5 Dimensionless groups

Drag coefficient $C_D \equiv \text{drag force}/\frac{1}{2}\rho V^2 A$

Discharge coefficient

$$C_d \equiv Q_{actual} \Bigg/ \left\{ A_{throat} \left\{ \frac{2\Delta p_{meter}/\rho}{1 - \left[\frac{A_{throat}}{A_{pipe}}\right]^2} \right\}^{\frac{1}{2}} \right\}$$

Fourier number	Fo	$\equiv (k/\rho c_p)t/L^2$
Froude number	Fr	$\equiv V/\sqrt{Lg}$
Grashof number	Gr	$\equiv g\beta\Delta T L^3 \rho^2/\mu^2$
Mach number	M	$\equiv V/a$
Nusselt number	Nu	$\equiv hL/k$
Prandtl number	Pr	$\equiv \mu c_p/k$
Reynolds' number - general	Re	$\equiv \rho VL/\mu$
" " - rotating disc	Re	$\equiv \rho\,\omega\,D^2/4\mu$
Weber number	We	$\equiv V(\rho\ell/\sigma)^{\frac{1}{2}}$
Pipeflow friction factor	f	$\equiv gDh_f/2\ell V^2$ (round pipes)
		$2gmh_f/\ell V^2$ (non-circular duct)
Wall shear stress coefficient	f	$\equiv \tau_w/\frac{1}{2}\rho V^2$

8.6 Composition of air

	Vol. Analysis	Grav. Analysis
Nitrogen (N_2 - 28.013)	0.7809	0.7553
Oxygen (O_2 - 31.999)	0.2095	0.2314
Argon (A_r - 39.948)	0.0093	0.0128
Carbon dioxide (CO_2 - 44.010)	0.0003	0.0005

Mean Molecular Weight M = 28.96

Specific Gas Constant R = 0.2871 kJ/(kgK)

8.7 Temperatures at the primary fixed points

Normal boiling point of oxygen (oxygen point)	$-182.97^\circ C$
Triple point of water	$0.01^\circ C$
Normal boiling point of water (steam point)	$100.00^\circ C$
Normal boiling point of sulphur (sulphur point)	$444.6^\circ C$
Normal melting point of silver (silver point)	$960.8^\circ C$
Normal melting point of gold (gold point)	$1063^\circ C$

8.8 Critical constants

	molecular weight	T_c(K)	P_c(bar) ($10^5 N/m^2$)	ρ_c(kg/m^3)
hydrogen	2.02	33.3	13.0	31
helium (4)	4.00	5.3	2.29	69.3
water vapour	18.02	647.30	221.2	318.3
nitrogen	28.01	126.1	33.9	311
oxygen	32.00	154.4	50.4	430
carbon dioxide	44.01	304.15	73.8	468

8.9 Approximate physical properties at 20°C, 1 bar ($10^5 N/m^2$)

	R $\frac{kJ}{kgK}$	ρ $\frac{kg}{m^3}$	c_p $\frac{kJ}{kgK}$	c_p/c_v	μ $\frac{mNs}{m^2}$=cP	k $\frac{W}{mK}$
hydrogen	4.16	0.082	14.3	1.40	8.8×10^{-3}	1.8×10^{-1}
helium	2.08	0.164	5.23	1.66	1.96×10^{-2}	1.4×10^{-1}
nitrogen	0.294	1.16	1.04	1.40	1.76×10^{-2}	2.6×10^{-2}
oxygen	0.260	1.31	0.91	1.40	2.03×10^{-2}	2.6×10^{-2}
carbon dioxide	0.190	1.80	0.84	1.28	1.47×10^{-2}	1.7×10^{-2}
air	0.287	1.19	1.005	1.40	1.82×10^{-2}	2.6×10^{-2}

(ii) Liquids

	ρ kg/m^3	c_p kJ/(kgK)	μ cP	k W/(mK)	σ N/m	β $10^{-3}K^{-1}$
water	1,000	4.19	1.002	0.6	0.073	0.21
mercury	13,600	0.14	1.55	8.7	0.51	0.18
castor oil	960	2.20	1000	0.18	0.039	
benzene	880	1.80	0.656	0.16	0.029	
ethyl alcohol	790	2.86	1.20	0.19	0.022	1.08
engine oil~	890	1.9	80	0.15	-	0.8
Freon 12	1,350	0.96	0.273	0.073	-	

(iii) Solids

	ρ kg/m^3	c_p kJ/(kgK)	k W/(mK)	α μm/(mK)
duralumin	2720	0.88	170	23
mild steel	7850	0.46	52	11
stainless steel (18% Ni, 8% Cr)	7810	0.46	16	18
brass (65/35)	8450	0.37	120	19
concrete	2400	0.88	1.1	10-14
wood (pine)	500	2.8	0.15	0.15
firebrick	170	0.81	0.38	3-9

(iv) Fuels

(a) Gases (fuels)

	Composition by Volume							Relative Density (Air) = 1	Calorific Value MJ/m^3 15°C 1.01325 bar		Theoretical Air
	N_2	H_2	CH_4	C_2H_6	C_3H_8	C_4H_{10}	C_3H_6		Gross	Net	Vol/Vol
Hydrogen		100						0.0696	12.10	10.22	2.38
Methane			100					0.5537	37.71	33.95	9.52
North Sea Gas	1.5		94.4	3.0	0.5	0.2		0.589	38.62	34.82	9.75
Propane*				1.5	91.0	2.5	5.0	1.523	93.87	86.43	23.76
Butane*			0.1	0.5	7.2	87.0	4.2	1.941	117.75	108.69	29.92

*Commercial Liquid petroleum Gas (L.P.G.) See also data on liquid fuels below.

(b) Liquids (fuels), typical values

	Composition % Mass			Density at 15°C kg/m^3	Calorific Value MJ/kg at 15°C	
	C	H	S		Gross	Net
Propane*	82.0	18.0		505	50.0	46.3
Butane*	81.9	17.0		575	49.3	45.8
Petrol	85.5	14.4	0.1	733	46.9	43.7
Kerosene	85.9	14.0	0.1	780	46.5	43.4
Diesel (Gas Oil)	85.7	13.4	0.9	840	45.4	42.4

8.10 Friction factor

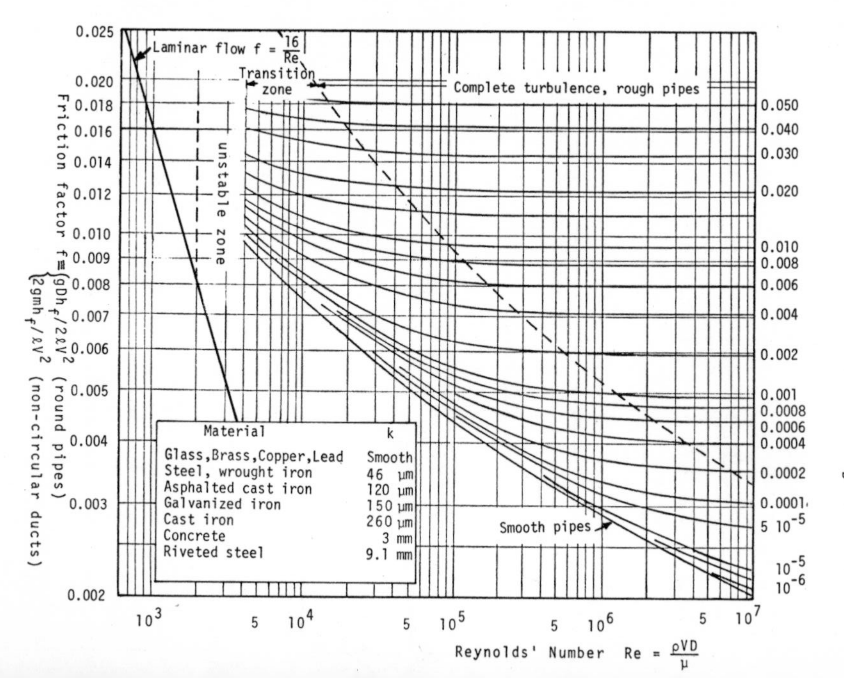

9. AUTOMATIC CONTROL

1. Block Diagrams

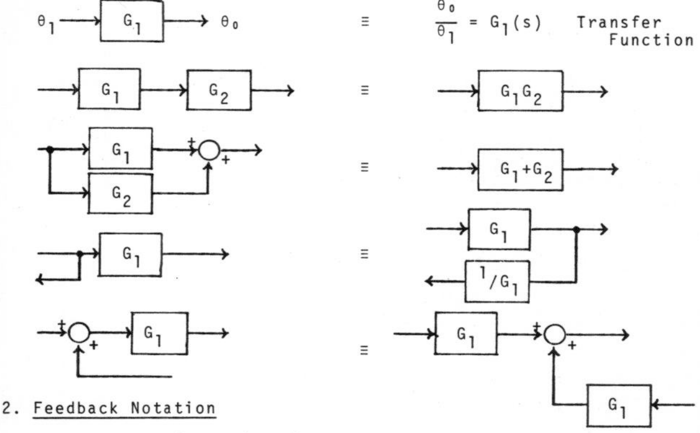

2. Feedback Notation

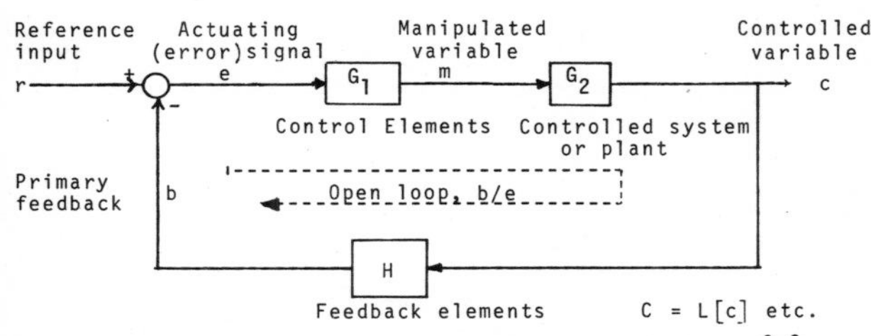

$C = L[c]$ etc.

Input/Output closed loop transfer function $= \frac{C}{R} = \frac{G_1 G_2}{1+G_1 G_2 H}$

Input/Error closed loop transfer function $= \frac{E}{R} = \frac{1}{1+G_1 G_2 H}$

Characteristic equation $= 1+G_1 G_2 H = 0$

$$G_1 G_2 H \triangleq KGH = \frac{K}{s^{\ell}} \frac{\prod_{i=1}^{m} (s+z_i)}{\prod_{i=1}^{n-\ell} (s+p_i)}$$

$-z_i, -p_i$ zeros and poles

ℓ system type number

n system order number

3. Stability Criteria for Linear Systems

3.1 Root location: No closed loop system pole may have positive real part

3.2 Routh Array

Characteristic equation $a_n s^n + a_{n-1} s^{n-1} \ldots + a_1 s + a_0 = 0$

	1	2	3	
n	a_n	a_{n-2}	a_{n-4}	"
n-1	a_{n-1}	a_{n-3}	a_{n-5}	"
n-2	b_1	b_2	b_3	"
n-3	c_1	c_2	c_3	"
"	"	"	"	"

$$b_1 = \frac{a_{n-1}a_{n-2} - a_n a_{n-3}}{a_{n-1}} \qquad b_2 = \frac{a_{n-1}a_{n-4} - a_n a_{n-5}}{a_{n-1}}$$

$$c_1 = \frac{b_1 a_{n-3} - a_{n-1} b_2}{b_1} \qquad c_2 = \frac{b_1 a_{n-5} - a_{n-1} b_3}{b_1} \qquad \text{etc.}$$

Number of closed loop poles with positive real part = number of sign changes in column 1.

3.3 Nyquist Encirclement

$$P = N + Z$$

N = number of clockwise encirclements of (-1,j0) by open loop locus

P = number of closed loop poles with positive real part

Z = number of open loop poles with positive real part

3.4 Gain Margin $= |KG(j\omega_g)H(j\omega_g)|^{-1}$, ω_g such that

$$\angle KG(j\omega_g)H(j\omega_g) = -180^o$$

3.5 Phase Margin $= 180^o + \angle KG(j\omega_p)H(j\omega_p)$, ω_p such that

$$|KG(j\omega_p)H(j\omega_p)| = 1$$

4. Rules of Root Locus Sketching

4.1 Every point, α, on the root locus for positive K satisfies

$$|G(\alpha)H(\alpha)| = 1/K$$

$$\angle G(\alpha)H(\alpha) = (1+2k)180^o \qquad k = 0, \pm1, \pm2.....$$

4.2 The number of branches of the root locus is equal to the number of poles.

4.3 Branches of the locus can be considered to start on the poles ($K = 0$) and terminate on zeros ($K = \infty$).

4.4 Points of the root locus exist on the real axis to the left of an odd number of poles plus zeros.

4.5 The locus is symmetrical with respect to the real axis.

4.6 The angles of asymptotes, α_k, to the root locus are given by

$$\alpha_k = \frac{\pm(2k+1)\pi}{n-m} \qquad k = 0, 1, 2.....$$

4.7 The intersection of the asymptotes and the real axis occurs at s_r

$$s_r = -\frac{\Sigma p_i - \Sigma z_i}{n-m}$$

4.8 The locus leaves the real axis or arrives at it at points α where α is given by

$$\frac{d}{d\alpha}\left[\ell n\{KG(\alpha)H(\alpha)\}\right] = 0$$

4.9 The intersection of the root locus and the imaginary axis can be found by application of Routh's Stability criteria.

$$\Delta p = \tfrac{1}{2}\rho(\Delta v^2 + \alpha u^2)$$

$$h_f = \frac{\Delta p}{\rho n} = 4f\frac{l}{d}\frac{\overline{u^2}}{2g} \qquad f = \frac{16}{R}$$

$$C_D = \frac{D}{q_0 C} = \frac{2\theta}{c} \qquad D_p = \int(p - p_0)\,ds\cos\theta \qquad D = \rho u_0^2\theta$$

Forced $v = \Omega r$ $\Gamma = 0$

Free $vr = const$ $\Gamma = (\omega_2 r_2^2 - \omega_1 r_1^2)k$

$$\Delta z = \frac{1}{2}\frac{\Omega^2}{g}\Delta r$$

$$F = \rho Q(u_4 - u_1) = \Delta p A \qquad Q = Au_2$$

$$u_2 = \tfrac{1}{2}(u_1 + u_4)$$

$$\eta_{prop} = \frac{u_1}{u_1 + \tfrac{1}{2}(u_4 - u_1)} \qquad \eta_{wind} = \frac{(u_1 + u_4)(u_1^2 - u_4^2)}{2u_1^3}$$

5. TABLE OF CHARACTERISTIC SYSTEMS

Typical Example			Basic form of transfer function G(s)
Electrical	Dynamic	Hydraulic	
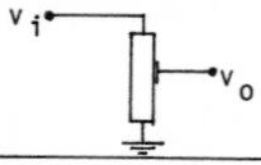 v_i, v_o	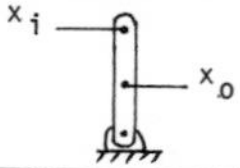 x_i, x_o	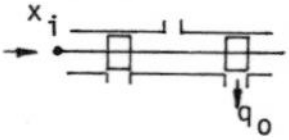 x_i, q_o	K
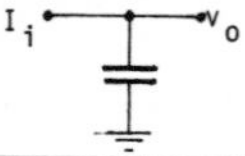 I_i, v_o	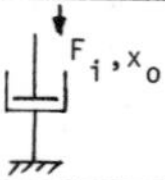 F_i, x_o	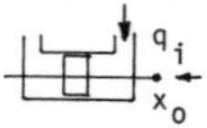 q_i, x_o	$\frac{1}{s}$
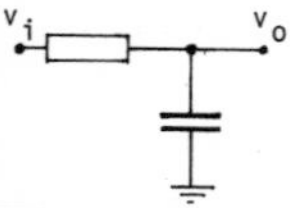 v_i, v_o	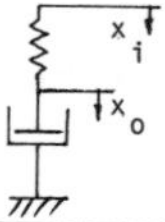 x_i, x_o	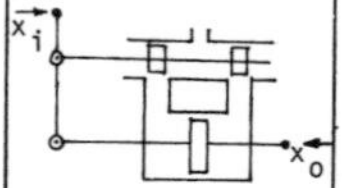 x_i, x_o	$\frac{1}{1+Ts}$
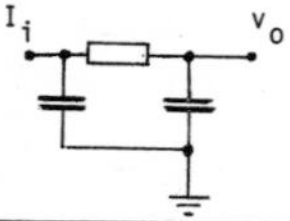 I_i, v_o	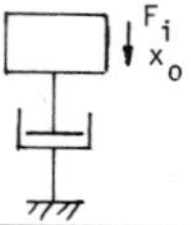 F_i, x_o	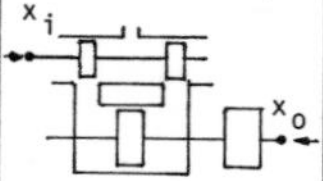 x_i, x_o	$\frac{1}{s(1+Ts)}$
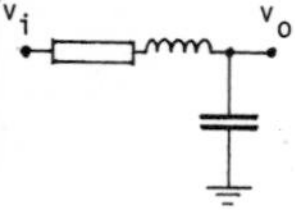 v_i, v_o	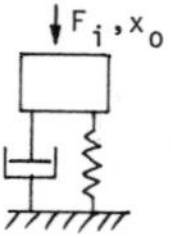 F_i, x_o	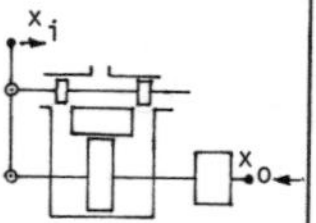 x_i, x_o	$\frac{\omega_n^2}{s^2+2\xi\omega_n s+\omega_n^2}$
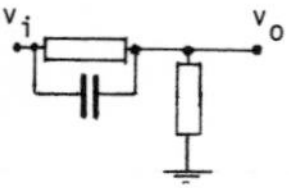 v_i, v_o	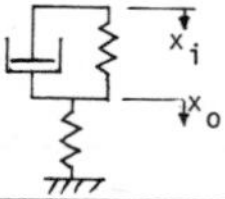 x_i, x_o	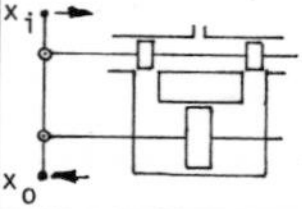 x_i, x_o	$\frac{1+\alpha Ts}{1+Ts}$
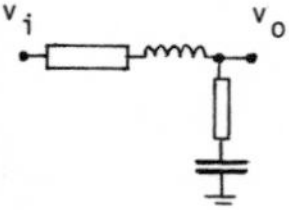 v_i, v_o	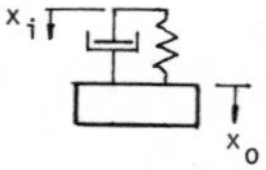 x_i, x_o	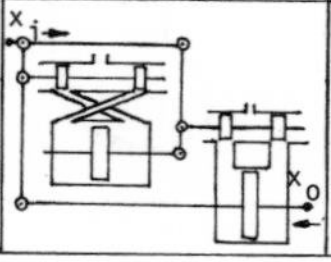 x_i, x_o	$\frac{\omega_n^2(1+2\xi s/\omega_n)}{s^2+2\xi\omega_n s+\omega_n^2}$

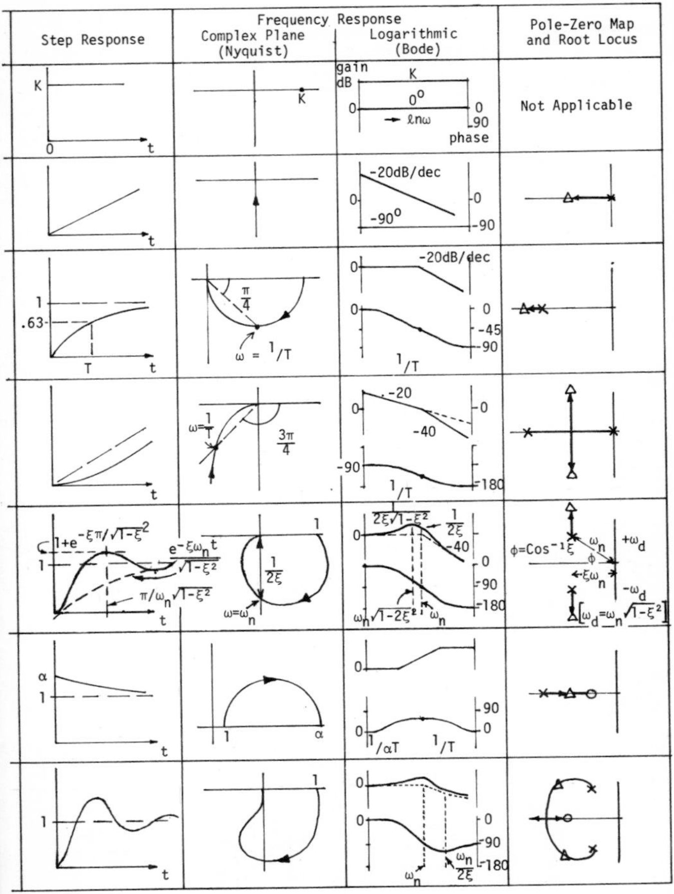

Step Response	Frequency Response Complex Plane (Nyquist)	Logarithmic (Bode)	Pole-Zero Map and Root Locus
K 0 t	K	gain dB K 0° 0 0 → $\ell n\omega$ -90 phase	Not Applicable
t		-20dB/dec 0 -90° 0 -90	
1 .63 T t	$\frac{\pi}{4}$ $\omega = {}^1/_T$	-20dB/dec 0 0 0 -45 -90 ${}^1/_T$	
t	$\omega=\frac{1}{T}$ $\frac{3\pi}{4}$	-20 0 0 -40 -90 -180 ${}^1/_T$	
$1+e^{-\xi\pi/\sqrt{1-\xi^2}}$ 1 $\frac{e^{-\xi\omega_n t}}{\sqrt{1-\xi^2}}$ $\pi/\omega_n\sqrt{1-\xi^2}$ t	1 $\frac{1}{2\xi}$ $\omega=\omega_n$	$2\xi\sqrt{1-\xi^2}$ $\frac{1}{2\xi}$ 0 -40 0 -90 -180 $\omega_n\sqrt{1-2\xi^2}$ ω_n	$\phi=\text{Cos}^{-1}\xi$ ω_n $+\omega_d$ ϕ $\xi\omega_n$ $-\omega_d$ $[\omega_d=\omega_n\sqrt{1-\xi^2}]$
α 1 t	1 α	0 90 0 0 ${}^1/_{\alpha T}$ ${}^1/_T$	
1 t	1	0 0 0 -90 -180 ω_n $\frac{\omega_n}{2\xi}$	

10. ELECTRICITY

Ohm's Law

$V = IR, \quad R = \frac{V}{I}, \quad I = \frac{V}{R}$

Power

DC Power $= VI = I^2R = V^2/R$

AC Power $= \mathrm{Re}(\underline{V}.\underline{I}) = |V||I|\mathrm{Cos}\phi$

Resistance

$I = \frac{a}{\rho_o(1+\alpha T)}\frac{dV}{dx} \qquad R = \int \frac{\rho_o(1+\alpha T)}{a}\,dx$

Inductance

$e = -L\frac{di}{dt} \qquad i = -\int \frac{V}{L}\,dt$

$L = N^2\mu_o\mu_r a/\ell$

for L R circuit decay $i = Ie^{-Rt/L}$

Stored energy $= \frac{1}{2}LI^2$

Capacitance

$Q = CV = \int i\,dt$

$i = \frac{dQ}{dt} = C\frac{dV}{dt}$

$C = \varepsilon_o\varepsilon_r(n-1)a/d$, for n parallel plates

$\varepsilon_o = 8.85.10^{-12}\ \mathrm{Fm}^{-1}$

for RC circuit discharge $i = -Ie^{-t/RC}$

Stored energy $= \frac{1}{2}CV^2$

$F = \frac{1}{2}\varepsilon_o\varepsilon_r a\left(\frac{V}{x}\right)^2$

Electrostatics

$F = \frac{Q_1Q_2}{4\pi\varepsilon_o\ r^2} \qquad \underline{F} = e\underline{E} = -e\ \mathrm{grad}V$

$Q = \oint \underline{D}.d\underline{S}(=\Psi) \qquad \underline{D} = \varepsilon_o\varepsilon_r\underline{E}$

Electromagnetism

$E = -N\frac{d\Phi}{dt} \qquad B = \frac{\mu_o I}{2\pi r}$

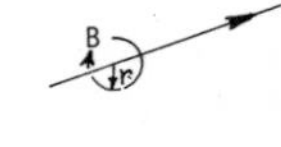

$F = B\ell I \qquad F = \frac{\mu_o I_1 I_2 \ell}{2\pi d}$

$\frac{dH}{d\ell} = \frac{I\mathrm{Sin}\alpha}{4\pi x^2}$

For solenoid $H = \frac{NI}{\ell}$

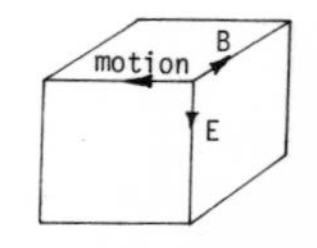

Magnetism

$$H = \frac{B}{\mu_o \mu_r}$$

For a magnetic circuit

$$B = \frac{\Phi}{a}$$

$$\Phi = \frac{NI}{\dfrac{\ell_1}{\mu_1 a_1} + \dfrac{\ell_2}{\mu_2 a_2}}$$

Stored Energy Density $= \frac{1}{2}HB = \frac{1}{2}\frac{B^2}{\mu_o}$

$$F = (\tfrac{1}{2}HB)a = \frac{B^2 a}{2\mu_o}$$

DC Machines

$$E = \frac{2Z}{c}\frac{n}{60}p\Phi \qquad T = \frac{I_a Z p \Phi}{2\pi c} \qquad \text{where } c = 2 \text{ (wave) or } 2p \text{ (lap)}$$

$$V = E \pm I_a R_a$$

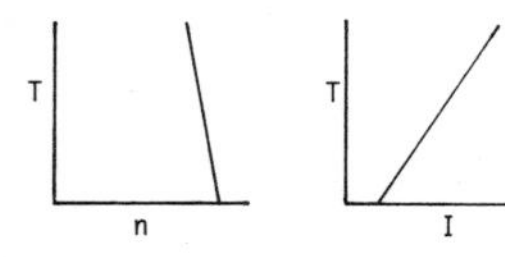

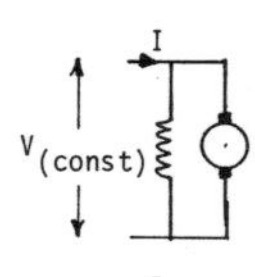

Shunt motor

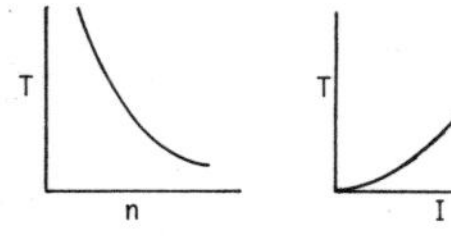

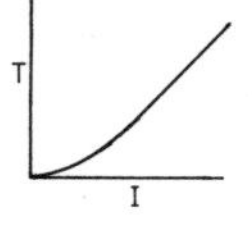

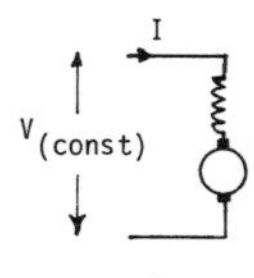

Series motor AC or DC

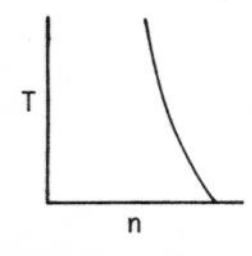

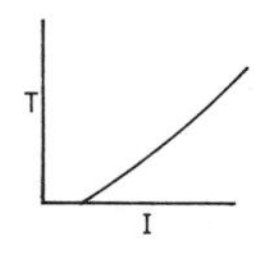

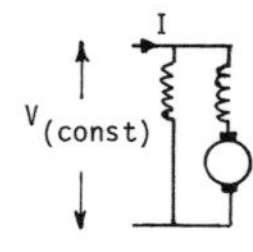

Compound motor

AC Machines

Synchronous speed = f/p

$E = 2.22\ k_d k_p Z f \Phi$(rms)

$$T \propto \frac{\Phi^2\ sR}{R^2 + (sX_o)^2}$$

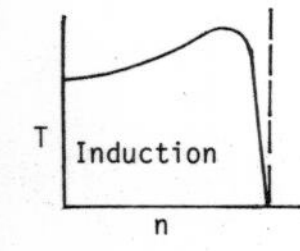

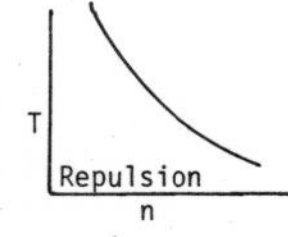

AC Circuits $V_{rms} = \frac{1}{\sqrt{2}} V_{max}$

Series LCR

$$Z = (R^2+(\omega L - \frac{1}{\omega C})^2)^{\frac{1}{2}} \qquad \omega = 2\pi f$$

$$\underline{Z} = R+j\omega L + \frac{1}{j\omega C} \qquad \cos\phi = \frac{R}{Z}$$

At resonance $\omega=\omega_o=\frac{1}{\sqrt{LC}}$ Q factor $= \omega_o\frac{L}{R}$

V_L, V_C, V_R, I, ϕ

Basic Op'Amp' Circuits

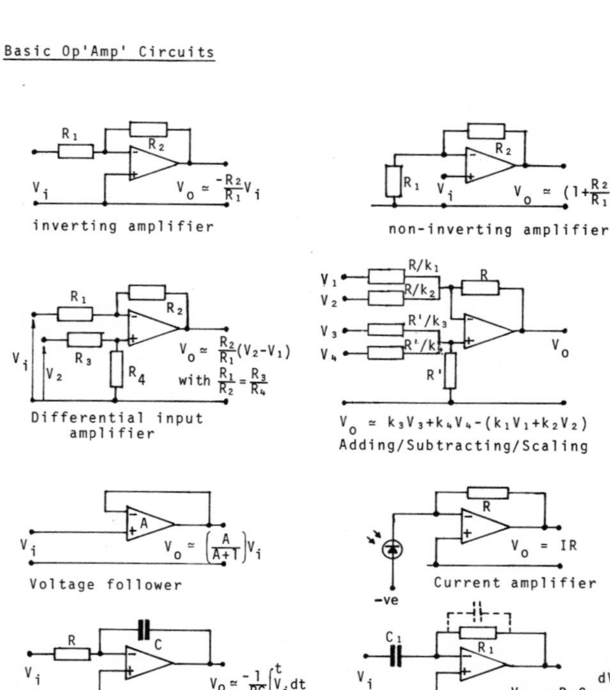

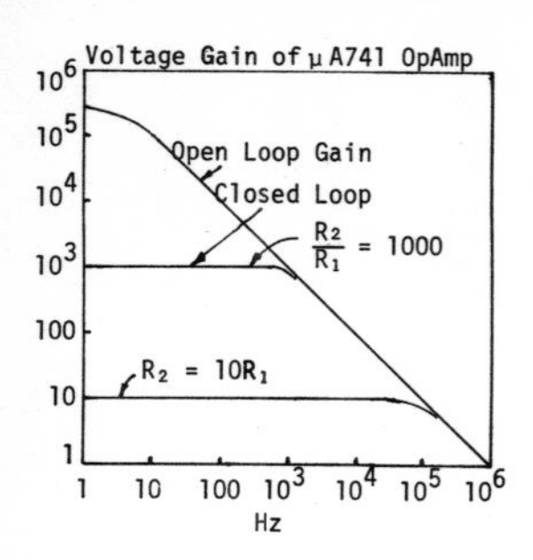

Colour Code

0 Black	2 Red	4 Yellow	6 Blue	8 Grey
1 Brown	3 Orange	5 Green	7 Violet	9 White

Preferred Values

10, 12, 15, 18, 22, 27, 33, 39, 47, 56, 68, 82

11 SOIL MECHANICS

11.1 Soil Classification

1. Size Classification

CLASSIFICATION		M.I.T. Size limits mm	B.S. Sieves used for separation mm
Gravel	Coarse		
		20	20
	Medium		
		6	6.3
	Fine		
		2	2
Sand	Coarse		
		0.6	0.6
	Medium		
		0.2	0.212
	Fine		
		0.06	0.063
Silt	Coarse		
		0.02	
	Medium		
		0.006	
	Fine		
		0.002	
Clay			

2. Casagrande Soil Classification, fine grained (50% or more passing B.S. No.200 sieve)

Silts and clays (Liquid limit less than 50)	Inorganic silts, silty or clayey fine sands, with slight plasticity	ML
	Inorganic clays, silty clays, sandy clays of low plasticity	CL
	Organic silts and organic silty clays of low plasticity	OL
Silts and clays (Liquid limit greater than 50)	Inorganic silts of high plasticity	MH
	Inorganic clays of high plasticity	CH
	Organic clays of high plasticity	OH
Highly organic soils	Peat and other highly organic soils	Pt

M silt
C clay
O organic

L low pasticity
H high plasticity

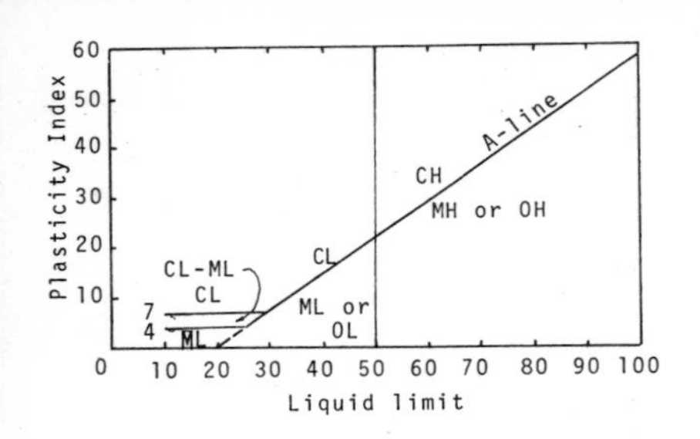

3. Volume-weight Relationships

Vol.		Weight
e	Air	0
Se	Water	$S e \gamma_w$
1	Solids	$G_s \gamma_w$

$$n = \frac{e}{1+e} \qquad e = \frac{n}{1-n}$$

$$w = \frac{Se}{G_s}$$

$$\gamma = \frac{G_s(1+w)}{1+e}\gamma_w$$

$$\gamma' = \frac{G_s-1}{1+e}\gamma_w = \gamma_{sat} - \gamma_w$$

4. Stratigraphic Table

Era	Period	Epoch
Cenozoic	Quaternary	Recent Pleistocene
	Tertiary	Pliocene Miocene Oligocene Eocene Paleocene
Mesozoic	Cretaceous Jurassic Triassic	
Paleozoic	Permian Carboniferous Devonian Silurian Ordovician Cambrian	
Precambrian		

Subdivisions of Quaternary

Relative Climate	U.K. Name
Warm (current)	Flandrian (Holocene)
Cold	Devensian
Warm	Ipswichian
Cold	Wolstonian
Warm	Hoxnian
Cold	Anglian
Warm	Cromerian
Cold	Beestonian
Warm	Pastonian
Cold	Baventian
Warm	Antian
Cold	Thurnian
	Ludhamian
	Waltonian

11.2 Stresses and Displacements in Elastic Half-space

1. Vertical stress at depth z below corner of uniformly loaded rectangle

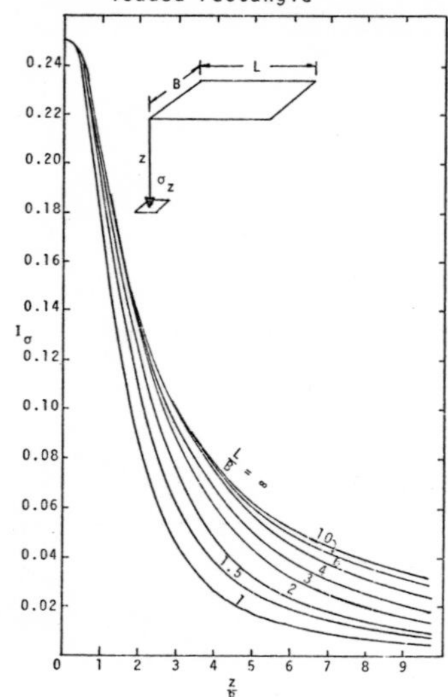

$$\sigma_z = qI_\sigma$$

2. Boussinesq

(a) Point load Q at surface

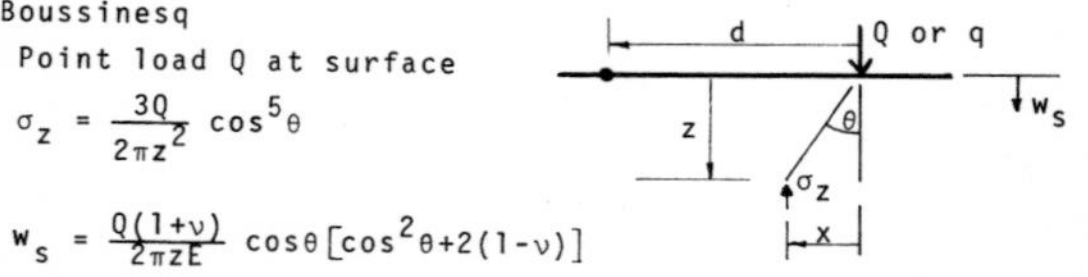

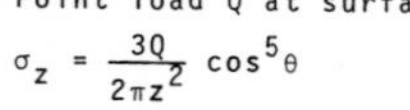

$$\sigma_z = \frac{3Q}{2\pi z^2}\cos^5\theta$$

$$w_s = \frac{Q(1+\nu)}{2\pi zE}\cos\theta\left[\cos^2\theta+2(1-\nu)\right]$$

(b) Line load q at surface

$$\sigma_z = \frac{2q}{\pi z}\cos^4\theta$$

$$w_s = \frac{2q(1-\nu^2)}{\pi E}\,\ell n\left(\frac{d}{x}\right)$$ where displacement at d is assumed = 0 $(d \geqslant x)$

3. Surface displacement of uniformly loaded rectangle

L/B Ratio	I_δ		
	Centre	Corner	Average
1	1.12	.56	.95
1.5	1.36	.68	1.15
2	1.53	.76	1.30
3	1.78	.89	1.53
4	1.96	.98	1.70
5	2.10	1.05	1.83
7	2.33	1.16	2.04
10	2.53	1.27	2.25
20	2.95	1.47	2.64
30	3.23	1.61	2.88
50	3.54	1.77	3.22
100	4.01	2.00	3.69
Circle	1.00	Edge .64	.85

$$w_s = qB\,\frac{1-\nu^2}{E}\,I_\delta$$

11.3 Terzaghi Bearing Capacity Factors

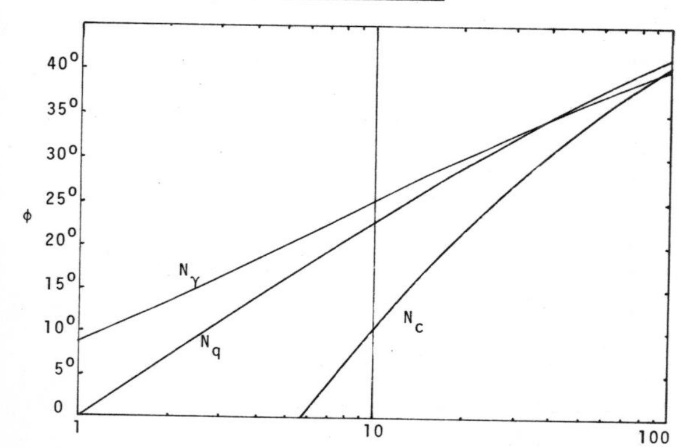

Strip footing $q = cN_c + \gamma DN_q + .5\gamma\, BN_\gamma$

Square footing $q = 1.3\, cN_c + \gamma DN_q + .4\gamma\, BN_\gamma$

Circular footing $q = 1.3\, cN_c + \gamma DN_q + .3\gamma\, BN_\gamma$

B = FOOTING WIDTH

Note: Reduce c and $\tan\phi$ to two thirds of measured values for local shear

11.4 Slope Stability

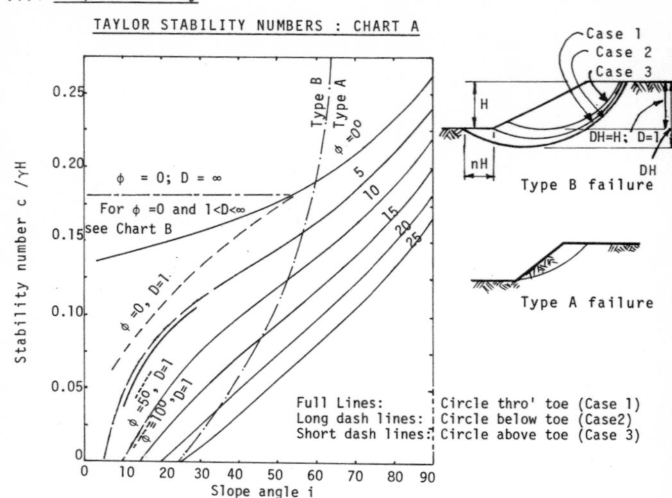

11.5 Consolidation-Time Curves

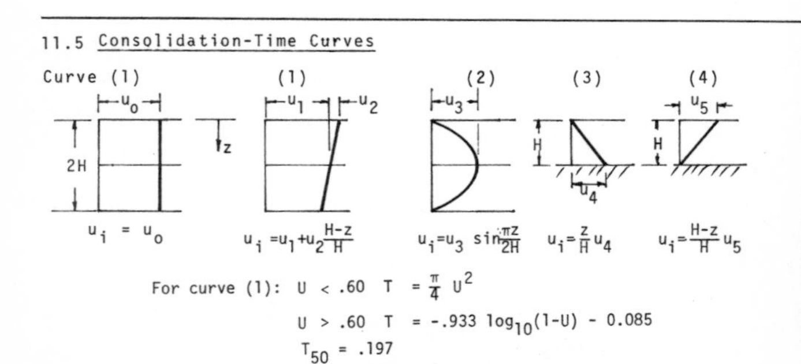

For curve (1): $U < .60 \quad T = \frac{\pi}{4} U^2$

$U > .60 \quad T = -.933 \log_{10}(1-U) - 0.085$

$T_{50} = .197$

$T_{90} = .848$

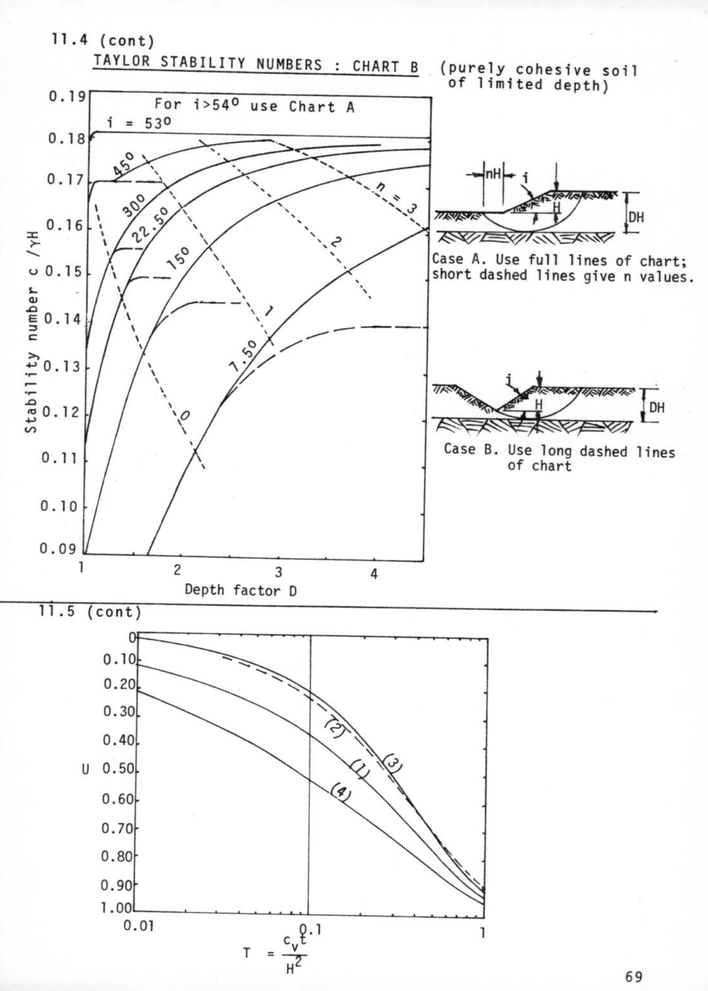

11.4 (cont)
TAYLOR STABILITY NUMBERS : CHART B
(purely cohesive soil of limited depth)
For i>54° use Chart A
i = 53°
45°
30°
22.5°
15°
7.5°
n = 3
2
1
0
Stability number c /γH
0.19
0.18
0.17
0.16
0.15
0.14
0.13
0.12
0.11
0.10
0.09
1
2
3
4
Depth factor D
nH
i
H
DH
Case A. Use full lines of chart; short dashed lines give n values.
i
H
DH
Case B. Use long dashed lines of chart
11.5 (cont)
0
0.10
0.20
0.30
0.40
U 0.50
0.60
0.70
0.80
0.90
1.00
0.01
0.1
1
(1)
(2)
(3)
(4)
T = c_v t / H^2

12. STRUCTURES

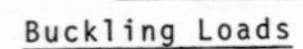

Buckling Loads

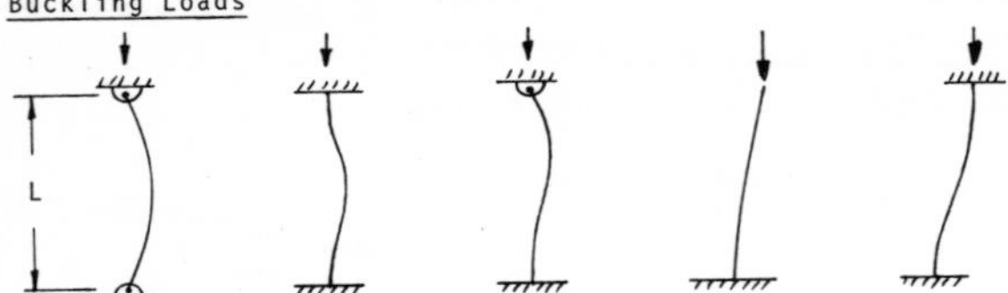

Buckling Load:	$\frac{\pi^2 EI}{L^2}$	$\frac{4\pi^2 EI}{L^2}$	$\frac{2.045\pi^2 EI}{L^2}$	$\frac{\pi^2 EI}{4L^2}$	$\frac{\pi^2 EI}{L^2}$
Effective Length:	L	0.5L	0.699L	2L	L

Beams bent about principal axis

ω is load/unit length	end slope	maximum deflection Δ
M; L	$\frac{ML}{EI}$	$\frac{ML^2}{2EI}$
W	$\frac{WL^2}{2EI}$	$\frac{WL^3}{3EI}$
ω	$\frac{\omega L^3}{6EI}$	$\frac{\omega L^4}{8EI}$
M; M	$\frac{ML}{2EI}$	$\frac{ML^2}{8EI}$
W; $\frac{L}{2}$; $\frac{L}{2}$	$\frac{WL^2}{16EI}$	$\frac{WL^3}{48EI}$
ω	$\frac{\omega L^3}{24EI}$	$\frac{5\omega L^4}{384EI}$
a; W; b; (Δ); $c = \sqrt{\frac{b(L+a)}{3}}$; $a \leqslant b$	$\theta_b = \frac{Wac^2}{2LEI}$; $\theta_a = \frac{L+b}{L+a} \cdot \theta_b$	$\frac{Wac^3}{3LEI}$; $a \leqslant b$

Fixed End Moments

LH end conditions moment	LH shear	Loading	RH shear	RH end conditions moment	maximum deflection Δ	maximum deflection position c
$\frac{\omega L^2}{12}$	$\frac{\omega L}{2}$	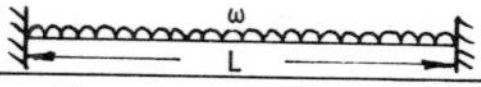 ω, L	$\frac{\omega L}{2}$	$\frac{\omega L^2}{12}$	$\frac{\omega L^4}{384EI}$	$\frac{L}{2}$
$\frac{WL}{8}$	$\frac{W}{2}$	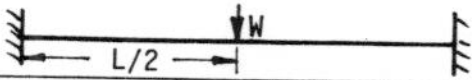 W, L/2	$\frac{W}{2}$	$\frac{WL}{8}$	$\frac{WL^3}{192EI}$	$\frac{L}{2}$
$\frac{Wab^2}{L^2}$	$\frac{Wb^2(L+2a)}{L^3}$	 W, a, b, c	$\frac{Wa^2(L+2b)}{L^3}$	$\frac{Wa^2b}{L^2}$	$\frac{2Wa^2b^3}{3EI(L+2b)^2}$ $a \leqslant b$	$\frac{2Lb}{L+2b}$
$\frac{6EI}{L^2}$	$\frac{12EI}{L^3}$	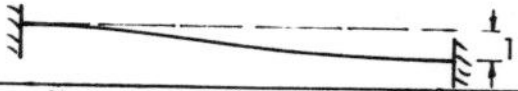 1	$\frac{12EI}{L^3}$	$\frac{6EI}{L^2}$	1	-
$\frac{Mb}{L^2}(2a-b)$	$\frac{M}{L}$	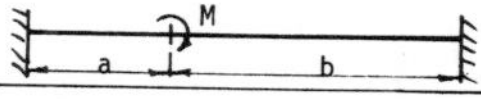 M, a, b	$\frac{M}{L}$	$\frac{Ma}{L^2}(2b-a)$	-	-
$\frac{\omega L^2}{30}$	$\frac{3\omega L}{20}$	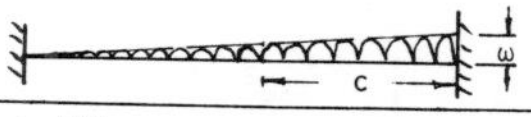 ω, c	$\frac{7\omega L}{20}$	$\frac{\omega L^2}{20}$	$\frac{\omega L^4}{764EI}$	0.475L
$\frac{3WL}{16}$	$\frac{11W}{16}$	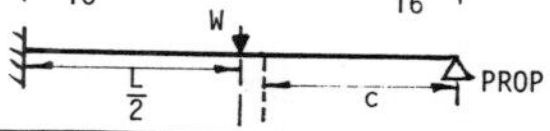 W, $\frac{L}{2}$, c, PROP	$\frac{5W}{16}$	0	$\frac{2WL^3}{215EI}$	0.447L
$\frac{Wab(L+b)}{2L^2}$	$\frac{Wb}{L}+\frac{M}{L}$	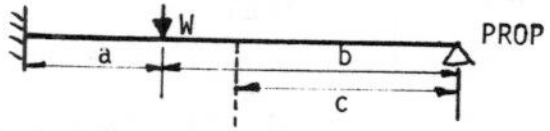 W, a, b, c, PROP	$\frac{Wa}{L}-\frac{M}{L}$	0	$\frac{Wa^2b}{6EI}\sqrt{\frac{b}{2L+b}}$ $b \geqslant 0.4142L$	$L\sqrt{\frac{b}{2L+b}}$ $b \geqslant 0.4142L$
$\frac{\omega L^2}{8}$	$\frac{5\omega L}{8}$	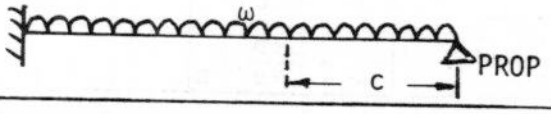 ω, c, PROP	$\frac{3\omega L}{8}$	0	$\frac{\omega L^4}{185EI}$	0.422L

Relations with elastic constants

$G = E/(2(1 + \nu))$ $\qquad$ $K = E/(3(1 - 2\nu))$

Simple bending $\frac{M}{I} = \frac{\sigma}{y} = \frac{E}{R}$

Torsion of circular section $\frac{T}{J} = \frac{\tau}{r} = \frac{G\theta}{L}$

Beam stiffness Coefficients

In the following the F's can be axial or shear forces, or, bending or torsional couples corresponding to the mode of deformation.

All beam and frame stiffness matrices may be built up from the following components of each beam element.

(a) axial stiffness [diagram: x_1, x_2, ℓ] giving $\begin{bmatrix} F_1 \\ F_2 \end{bmatrix} = \frac{EA}{\ell} \begin{bmatrix} 1 & -1 \\ -1 & 1 \end{bmatrix} \begin{bmatrix} x_1 \\ x_2 \end{bmatrix}$

(b) torsional stiffness [diagram: x_1, x_2] giving $\begin{bmatrix} F_1 \\ F_2 \end{bmatrix} = \frac{GJ}{\ell} \begin{bmatrix} 1 & -1 \\ -1 & 1 \end{bmatrix} \begin{bmatrix} x_1 \\ x_2 \end{bmatrix}$

(c) bending stiffness and lateral deflection stiffness in one plane

[diagram: x_3, x_4, x_1, x_2] giving

$$\begin{bmatrix} F_1 \\ F_2 \\ F_3 \\ F_4 \end{bmatrix} = \frac{EI}{\ell^3} \begin{bmatrix} 12 & -12 & 6\ell & 6\ell \\ -12 & 12 & -6\ell & -6\ell \\ 6\ell & -6\ell & 4\ell^2 & 2\ell^2 \\ 6\ell & -6\ell & 2\ell^2 & 4\ell^2 \end{bmatrix} \begin{bmatrix} x_1 \\ x_2 \\ x_3 \\ x_4 \end{bmatrix}$$

In each case if one end is fixed and considered as a reaction. its deflections and forces may be ignored with a corresponding reduction of the stiffness matrix. Another possible form of reaction for case (c) occurs if the reaction end is pinned. Then the stiffness matrix components for the other end are given by

(c)(i) [diagram: x_2, x_1, Pin] giving $\begin{bmatrix} F_1 \\ F_2 \end{bmatrix} = \frac{EI}{\ell^3} \begin{bmatrix} 3 & 3\ell \\ 3\ell & 3\ell^2 \end{bmatrix} \begin{bmatrix} x_1 \\ x_2 \end{bmatrix}$

or

(c)(ii) [diagram: Pin, x_2, x_1] giving $\begin{bmatrix} F_1 \\ F_2 \end{bmatrix} = \frac{EI}{\ell^2} \begin{bmatrix} 3 & -3\ell \\ -3\ell & 3\ell^2 \end{bmatrix} \begin{bmatrix} x_1 \\ x_2 \end{bmatrix}$

A general plane frame element will have components (a) and (c) and require a 6 x 6 stiffness matrix. A general space frame element will require components of (a), (b) and (c) - the latter twice for two planes of bending - and will require a 12 x 12 stiffness matrix. The three modes of deflection (a), (b), (c) are orthogonal and may be combined into larger matrices with 0's in all unspecified positions. Space frame elements will, in general, have different values of I in the two principal planes of bending.

Shear

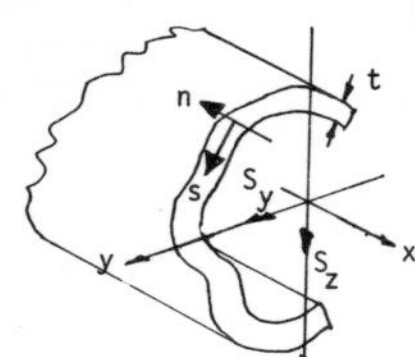

Shear flow per unit length of wall resulting from the applied shear forces S_z, S_y is

$$q = t\tau_{xs} = \frac{(-S_z)}{I_{yy}I_{zz}-(I_{yz})^2}\left(I_{yy}\int_A zdA - I_{yz}\int_A ydA\right) + \frac{(-S_y)}{I_{yy}I_{zz}-(I_{yz})^2}\left(I_{zz}\int_A ydA - I_{yz}\int_A zdA\right)$$

The resultant force from this shear flow acts through the SHEAR CENTRE.

Torsion

For a circular section $\frac{T_x}{J} = \frac{\tau_{\theta x}}{r} = \frac{G\theta}{L}$

$J = \frac{\pi D^4}{32}$ for a solid section

$= \frac{\pi}{32}\left(D^4_{outer} - D^4_{inner}\right)$ for a hollow section

For a thin walled closed section

$$T_x = 2Aq = \frac{4A^2G}{\int_0^s \frac{ds}{t}} \cdot \frac{\theta}{L}$$

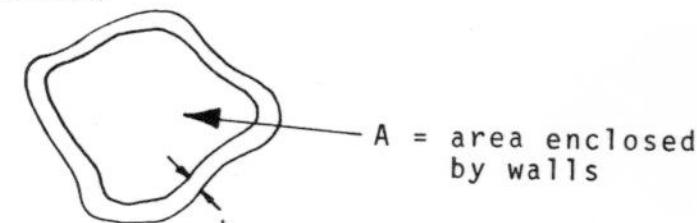

For a thin rectangular section

$$T_x = \frac{dt^2}{3}\tau_{zx\,max} = \frac{dt^3G}{3}\frac{\theta}{L}$$

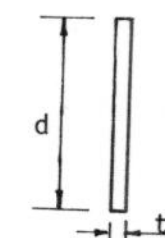

Asymmetric Bending

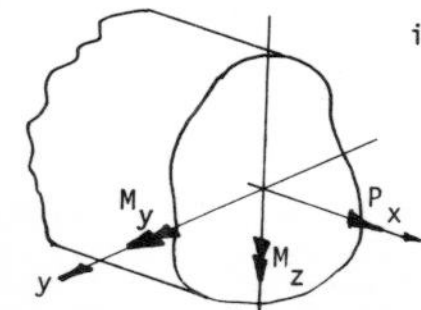

in terms of general axes

$$\sigma_{xx} = \frac{P_x}{A} + \frac{M_y(zI_{zz}-yI_{yz})}{I_{yy}I_{zz}-(I_{yz})^2} - \frac{M_z(yI_{yy}-zI_{yz})}{I_{yy}I_{zz}-(I_{yz})^2}$$

When the principal axes m,n lie in directions y,z, then:

$$\sigma_{xx} = \frac{P}{A} + \frac{nM_m}{I_{mm}} - \frac{mM_n}{I_{nn}}$$

Stress and strain transformations

Mohr circle of stresses

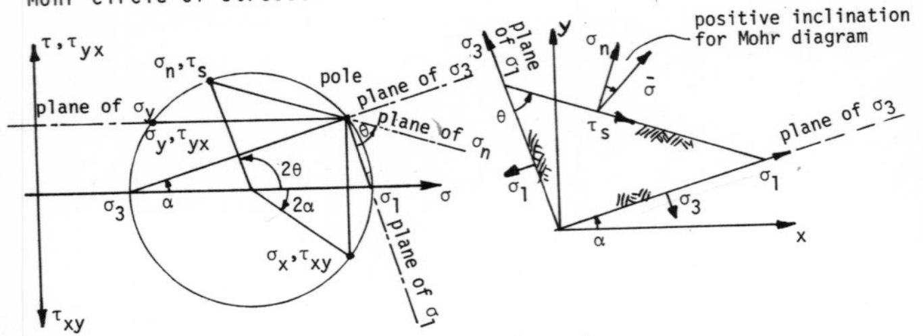

Equilibrium of prism in σ_1, σ_3 directions gives

$$\sigma_1 \cos\theta = \sigma_n \cos\theta + \tau_s \sin\theta$$
$$\sigma_3 \sin\theta = \sigma_n \sin\theta - \tau_s \cos\theta$$

Then: $\sigma_1 \cos^2\theta + \sigma_3 \sin^2\theta = \sigma_n = \frac{\sigma_1}{2}(1+\cos 2\theta) + \frac{\sigma_3}{2}(1-\cos 2\theta)$

$$= \frac{\sigma_1+\sigma_3}{2} + \frac{\sigma_1-\sigma_3}{2}\cos 2\theta$$

$$(\sigma_1-\sigma_3)\sin\theta\cos\theta = \tau_s = \frac{\sigma_1-\sigma_3}{2}\sin 2\theta$$

Also: $\sigma_1 = \frac{\sigma_x+\sigma_y}{2} + \tau_{max}$; $\sigma_3 = \frac{\sigma_x+\sigma_y}{2} - \tau_{max}$; $\tau_{max} = \sqrt{\left(\frac{\sigma_x-\sigma_y}{2}\right)^2 + (\tau_{xy})^2}$

and: $$\tan 2\alpha = \frac{-2\tau_{xy}}{\sigma_x-\sigma_y}$$

Two-dimensional strain system

$\varepsilon_x, \varepsilon_y, \varepsilon$ are direct strains 'corresponding' to $\sigma_x, \sigma_y, \sigma$

$\frac{\gamma_{xy}}{2}, \frac{\gamma}{2}$ are shear strains 'corresponding' to τ_{xy}, τ

Three-dimensional stress system

If the principal stresses are σ_1, σ_2, σ_3, the principal shear stresses are $(\sigma_1 - \sigma_2)/2$, $(\sigma_2 - \sigma_3)/2$ and $(\sigma_3 - \sigma_1)/2$.

Strain energy per unit volume U may be expressed as

$$U = (\sigma_1 + \sigma_2 + \sigma_3)^2/18K$$
$$+\left[(\sigma_1 - \sigma_2)^2 + (\sigma_2 - \sigma_3)^2 + (\sigma_3 - \sigma_1)^2\right]/12G$$

13. SYMBOLS INDEX

GREEK ALPHABET

Α, α alpha	Η, η eta	Ν, ν nu	Τ, τ tau
Β, β beta	Θ, θ theta	Ξ, ξ xi	Υ, υ upsilon
Γ, γ gamma	Ι, ι iota	Ο, ο omicron	Φ, φ phi
Δ, δ delta	Κ, κ kappa	Π, π pi	Χ, χ chi
Ε, ε epsilon	Λ, λ lambda	Ρ, ρ rho	Ψ, ψ psi
Ζ, ζ zeta	Μ, μ mu	Σ, σ sigma	Ω, ω omega

MATHEMATICAL SYMBOLS

L[] - Laplace Transform
$\underline{\Delta}$ - defined as
Σ - repeated summation
Π - repeated multiplication
∂ - partial differential
|a| - modulus
∇ - Laplace differential, Del, Nabla
$\underline{a}$ - vector
$\hat{\underline{a}}$ - unit vector
⊥ - 'at right angles to'
. - scalar (dot) product
x - vector (cross) product
Re - real part of complex number
Im - imaginary part of complex number

Symbol	Page No. of use		Recommended S.I. Unit
a	51	velocity of sound	m/s
a	38	lattice parameter	m
a	41	½ crack length	m
a	60	area	m^2
a	37	acceleration	m/s^2
A	48,50	area	m^2
A	38	Atomic weight	-
A	47	availability function (non-flow)	kJ
B	47	availability function (flow)	kJ
B	60	magnetic flux density	T
B	66,67	breadth of footing	m

* of specified fluid

c	67	cohesion	kN/m^2
c	49	velocity	m/s
c_p, c_v	47	specific heat	kJ/(kgK)
c_v	69	coefficient of consolidation	m^2/s
C	60	capacitance	F
C	39	concentration	mol/m^2
C	50	Chézy coefficient	$(m^{1/2})/s$
C_d	50,51	discharge coefficient	-
C_d	51	drag coefficient	-
d	38	interatomic spacing	m
$d()$	38	interplanar spacing	m
d	50	depth of flow	m
d_1, d_2	50	" " " before jump, after jump	m
D	51,54	diameter	m
D	39	diffusion coefficient	m^2/s
D	67	depth of overburden	m
D	68,69	depth factor	-
e	65	void ratio	-
E	41,70	Young's Modulus	N/m^2
ΔE	38	energy difference	J
f	61	supply frequency	Hz
f	54	friction factor	-
f	51	wall shear stress coefficient	-
f	47	specific Helmholtz free energy function	kJ/kg
F	47	Helmholtz free energy function	kJ
F, F_1	50,51	Froude Number, before jump	-
F	60	force	N
g	47	specific Gibbs free energy function	kJ/kg
G_s	65	specific gravity of solids	-
G	47	Gibbs free energy function	kJ
G	41,72	modulus of rigidity	N/m^2
G_1, G_2, G	55	transfer function	-
h_f	51,54	frictional head loss	m*
h	47	specific enthalpy	kJ/kg
h	51	heat transfer coefficient	$W/(m^2K)$
H	47	enthalpy	kJ
H	60	magnetising force magnetic field strength	A/m
H	55	transfer function	-
I	60	current	A
I	35,70	second moment of area	m^4
I	35	moment of inertia	kgm^2
J	39	diffusion flux	$kg/(m^2s)$
k	48,51	thermal conductivity	W/(mK)
k	54	surface roughness	µm
k	35,49	radius of gyration	m
k_d, k_p	61	distribution factor, pitch factor of winding	-
K_p	55	gain constant	-
K	41	stress intensity factor	$MNm^{-3/2}$
K	41,72	bulk modulus	N/m^2
ΔK	41	stress intensity range	$MNn^{-3/2}$
K_p	47	equilibrium constant	$(atmos)^{1/n}$
ℓ	51	length	m
ℓ	60	length of conductor	m
L	51	characteristic length	m
L	60	inductance	H
m	47	mass	kg
m	51,54	mean hydraulic radius	m
$\dot{m}$	47,50	mass flowrate	kg/s

M	70	moment	Nm
M	51	Mach number	-
n	65	porosity	-
n	61	speed of rotation	r/min
n	50	Manning roughness coefficient	-
n	38	atoms per unit cell	-
n_i	41	total cycles at stress amplitude	-
N_p	41	total cycles to failure at strain amplitude	-
N_i	41	" " " " " stress "	-
N	60,61	number of turns	-
N	7,38	Avogadro's number	kg/(kgmol)
N_c	47	cycles per unit time	Hz
N'	39	total number of atoms	-
p	30	probability	-
p_m	47	mean effective pressure	N/m^2
p	61	number of pole pairs	-
P	47	thermodynamic probability or No.of quantum states	-
P_i	47	engine indicated power	W
q	66,67	surface normal stress	kN/m^2
q	48	heat flowrate per unit area emissive power	W/m^2
q_b	48	emissive power for black body	W/m^2
Q	39	activation energy	J
Q	47	heat (input + ve)	kJ
Q	49	volumetric flowrate	m^3/s
Q	41	crack shape factor	-
Q	60	charge	C
r	41	r.m.s. of stress	MN/m^2
R	60,61	resistance, resistance per phase	Ω
R	50	hydraulic radius	m
R	47	characteristic gas constant	kJ/(kgK)
R_o	47	Universal gas constant	kJ/(kgmolK)
s	47	specific entropy	kJ/(kgK)
s	61	fractional slip	-
S	65	degree of saturation	-
S	47	entropy	kJ/K
S	50	channel slope in uniform flow	-
S_f	50	friction slope	-
S_o	50	invert slope	-
t	37	time	s
T	68	time factor	-
T	61	torque	Nm
T	50	water surface width	m
T	47	temperature (absolute)	K
ΔT	48	temperature difference	°C
T_g	43	glass transition temperature	K
u	47	specific internal energy	kJ/kg
u	37	velocity	m/s
u	68	pore pressure	kN/m^2
U	68	degree of consolidation	-
U	47	internal energy	kJ
v	47	specific volume	m^3/kg
v	37	velocity	m/s
v_o	47	molar volume	$m^3/(kgmol)$
V	60	voltage	V
V	38,49	volume	m^3
V	38	volume of unit cell	m^3
V	50,51	velocity	m/s
V_s	47	cylinder swept volume	m^3

w	65	water content	-
w_s	66,67	surface displacement	m
W	47	work (output,+ve)	kJ
X_o	61	leakage reactance per phase	Ω
Y	49	change in specific energy through a machine	J/kg
z	49	potential head	m*
Z	61	number of armature conductors	-
Z	62	impedance	Ω
α	41	coefficient of linear expansion	μm/(mK)
α	60	resistance coefficient	Ω/K
β	51	coefficient of volumetric expansion	K^{-1}
γ	65	unit weight of soil	kN/m^3
γ	47,50	specific heat ratio	-
γ_w	65	unit weight of water	kN/m^3
γ_{sat}	65	unit weight of saturated soil	kN/m^3
γ'	65	submerged unit weight of soil	kN/m^3
ε	54	relative roughness	-
$\varepsilon_o, \varepsilon_r$	60	permittivity, free space, relative	F/m,-
ε	48	emissivity	-
ε_p	41	plastic straining range	μm/m
η	49	viscosity (dynamic)	mNs/m^2
μ_o, μ_r	60	permeability of free space, relative	H/m, -
μ	49,51	dynamic viscosity	mNs/m^2
μ	37	coefficient of friction	-
ν	72,41,66,67	Poisson's ratio	-
ξ	58,23	damping ratio	-
ρ_o	60	resistivity	Ωm
ρ	41,42,43,51	density	kg/m^3
σ	48	Stefan-Boltzmann constant	$W/(m^2K^4)$
σ	51	surface tension in contact with air	N/m
σ_y	41	proof or yield stress	N/m^2
σ_f	41	ultimate (failure) stress	N/m^2
τ	49,51	shear stress	N/m^2
ϕ	67	friction angle of soil	o
Φ	60,61	magnetic flux, flux per pole	Wb
ω	70,71	load per unit length	N/m
ω	51	angular velocity	rad/s
ω_n	58	natural frequency	rad/s
ω_o	23,62	natural frequency	rad/s
ω_d	59	damped natural frequency	rad/s

14. KEYWORD INDEX

A. AC circuits 62
AC machines 61
Acceleration equations (constant) 37
Acceleration of fluids 49
Activation energy 39
Air, composition 51
Algebra 13,19
Amplifier arrangements 62
Area of shapes 35,36
Asymmetric bending 73
Atomic number 46
Atomic sizes 38
Atomic volume 42
Atomic weight 46
Availability function 47
Avogadro's number 7,38

B. BASIC Language 8
Beams 70
Beam deflections 70,71
Beam stiffness coefficients 72
Bearings, Reynolds' equation 49
Bending 70,71
Bernoulli's equation 49
Binomial distribution 30
Binomial series 15
Black body 48
Block diagrams (control) 55
Bode diagram 59
Bohr magneton 7
Boiling point 51
Boltzmann's constant 7
Bonding 38,42

Boussinesq relationships 66
Buckling loads 70
Bulk modulus 41,72
C. Calorific value 53
Capacitance 6,60
Carnot efficiency 47
Casagrande soil classification 64
Cavitation number 50
Centre of buoyancy 49
Centre of gravity 35,49
Centre of mass 35
Centre of pressure 49
Centroid 35,49
Channels 50
Characteristic equation (control) 55,56
Characteristic gas constant 47
Chézy equation 50
Coefficient: Linear expansion 41,42,43
  Resistance 42
Cohesive energy 42
Colour codes (resistors) 63
Complex numbers 22
Condon-Morse equation 38
Conductivity 6
Conjugate depths 50
Consolidation, degree 68
Consolidation - Time curves 68
Constants 7
Continuity equation 47,50
Control: Order number 55
  Type number 55
Control systems: Dynamic 58
  Electrical 58
  Hydraulic 58
Convection 48
Conversion factors 4,5,6
Coriolis 37
Cosine rule for triangle 20
Critical constants, pressure, temperature, density 51
Critical damping 23
Cross product 13
Crystallography 38
Crystal structure 42
Cubic crystals 38,39
Cumulative distribution function 31,32
Curl 14
Curve fitting 28,29
Cylindrical coordinates 14
Damping ratio 23,58
DC machines 61
Defects 39
Definite integrals 26
Degrees of freedom 33
Density 4,38,41,42,43,53
Depth factors 68,69
Differentials 24
Differential equations 22
Differentiation, rules 24
Diffusivity 39
Dimensionless groups 51
Discharge coefficient 50,51
Div 14
Dot product 13
Drag coefficient 51
Dynamics, fluids 49
Dynamic responses 23
E. Elastic constants 41,43,72
Elastic half spaces 65
Electricity, formulae 60
Electricity, units 2,6,60
Electromagnetism 60
Electron: Charge 7
  Rest mass 7
  Charge/Mass ratio 7
Electrostatics 60
Elements 44,46
Emissivity 48
Energy gap 42
Engine indicated power 47
Enthalpy 47
Entropy 47
Equations: 1st Order Differential 22
  2nd " " 23
Errors 33,34
Error function 26,39,40
Euler buckling 70
Euler's formula 49
Exact DE 22
Expansion coefficient 41,42,53
Experimental samples 33
F. Faraday constant 7
Fatigue 41
Fermi energy 42
Feedback notation 55
Fick's Laws 39
Finite difference formulae 28
First Law 47
Fixed end moments 71
Flow coefficient 49
Flowrate of gases 50
Flowrate units 4
Fluid mechanics 47,49
Footings 67
Forced convection 48
Forced oscillation 23
Forward path 55
Fourier number 51
Fourier series 16
Fracture stress 41
Frequency response 23,59,63
Friction angle of soil 67
Friction coefficients 37
Friction factor 51,54
Friction head loss 51,54
Froude number 50,51
Fuels 53
G. Gas properties 52,53
Gas flow 50
Gauss' theorem 60
General normal distribution 32
Geological divisions 65

Gibbs function 47
Glass transition temperature 43
Grad 14
Grashof number 51
Gravitational constant 7
Greek alphabet 75
H. Hall mobility 42
Head coefficient 49
Heat conduction 48
Heat transfer 48
Heat transfer coefficient 48
Helmholtz function 47
Homogenous DE 22
Hyperbolic relations 19
I. Indefinite integrals 24
Identities 15
Inductance 6,60
Instrument error 34
Integrals 24
Integration by parts 24
Internal energy 47
International atmosphere 5
Interplanar spacing 38
Ionic bond equation 38
Ionic radii 42
L. Lagrange's interpolation formula 29
Laminar flow 54
Laplace 15
Laplace transforms 27
Lattice constant (parameter) 38,42
Least-squares fitting 28
Linear DE 22,23
Liquid limit 65
Liquids, properties 52,53
Loads: Line 66
Point 66
Logarithmic mean temperature difference 4
Logarithmic relations 21
Loops: Closed 55
Open 55
M. Mach number 51
Machines, electrical: DC 61,
AC 61
Machines, hydraulic 49
Maclaurin's series 16
Magnetism, formulae 60,61
Magnetism, units 2,6
Manning equation 50
Manson-Coffin Law 41
Margin (control): Gain 56
Phase 56
Materials 38,52,53
Mathematical symbols 75
Maxwell relationships 48
Melting point 42,51
Mesh openings 64
Metacentre 49
Miller index system 38,39
Miner's rule 41
Modulus of rigidity 41,72
Mohr's Circle 37
Molar volume 47
Molecular weight 52
Moments of beams 71
Moments of inertia 35
Most probable error 34
Motors, electrical 61
N. Natural frequency 23,58,62
Napier's rule 20
Newtonian flow 49
Newton's method 28
Normal distribution 31,32
Nozzle flow 50
Numerical analysis 28
Numerical integration 29
Nusselt number 51
Nyquist 56,59
O. Ohm's Law 60
Open channels 50
Op 'Amp': Circuits 62
Characteristics 63
Orbitals 45
Oscillations 23,59
P. Parallel axes theorem 35
Paris equation 41
Partial differentiation 22,27
Pascal's triangle 30
Perfect gas equations 47,50
Periodic table of the elements 44
Perpendicular axes theorem 35
Planck's constant 7
Plasticity index (soil) 65
Phase transformations 38
Physical properties of solids 41,53
Liquids and gases 52,53
Pipe flow friction factor 51,54
Poisson distribution 30
Poisson's ratio 41,72
Polar moment 35
Poles 55
Pole-Zero map 59
Polymer structure 43
Population variance 33
Porosity 65
Potential head 49
Prandtl number 51
Primary fixed points 51
Principal shell 45
Principal stresses 74
Probability 30
Probability density function 31
Proof stress 41
Q. Q factor 62
Quadratic equation, solution 21
R. Radiation 48
Radius of gyration 35,49
Random variables 30
Rayleigh distribution 4
Rectification 19
Resistance: Formulae 60
Preferred values 63
Colour code 63